Educational Import

Educational Import

Local Encounters with Global Forces in Mongolia

*Gita Steiner-Khamsi and
Ines Stolpe*

EDUCATIONAL IMPORT
© Gita Steiner-Khamsi and Ines Stolpe, 2006.
Cover photograph © G. Steiner-Khamsi.
3.1 © D. Amgalan.
6.1 © I. Stolpe and G. Steiner-Khamsi.

All rights reserved. No part of this book may be used or reproduced in any manner whatsoever without written permission except in the case of brief quotations embodied in critical articles or reviews.

First published in 2006 by
PALGRAVE MACMILLAN™
175 Fifth Avenue, New York, N.Y. 10010 and
Houndmills, Basingstoke, Hampshire, England RG21 6XS
Companies and representatives throughout the world.

PALGRAVE MACMILLAN is the global academic imprint of the Palgrave Macmillan division of St. Martin's Press, LLC and of Palgrave Macmillan Ltd. Macmillan® is a registered trademark in the United States, United Kingdom and other countries. Palgrave is a registered trademark in the European Union and other countries.

ISBN 978-1-4039-6810-4

Library of Congress Cataloging-in-Publication Data

Steiner-Khamsi, Gita.
   Educational import : local encounters with global forces in Mongolia / Gita Steiner-Khamsi and Ines Stolpe.
     p. cm.
   Includes bibliographical references and index.

   1. Education—Mongolia. 2. Educational change—Mongolia. I. Stolpe, Ines. II. Title.
LA1341.S74 2006
370.95173—dc22   2005053509

A catalogue record for this book is available from the British Library.

Design by Newgen Imaging Systems (P) Ltd., Chennai, India.

First edition: April 2006

10 9 8 7 6 5 4 3 2 1

Transferred to Digital Printing in 2013

Contents

List of Tables, Figures, Maps, and Photographs   vi
Acknowledgments   vii

1. Going Global: Studying Late Adopters of Traveling Reforms   1
2. Educational Import in Mongolia: A Historical Perspective   23
3. Bypassing Capitalism   51
4. Exchanging Allies: From Internationalist to International Cooperation   67
5. Structural Adjustment Reforms, Ten Years Later   85
6. The Mongolization of Student-Centered Learning   109
7. Outcomes-Based Education: Banking on Policy Import   131
8. Speaking the Language of the New Allies with the Voucher (Non-) Reform   147
9. What if There is Nothing to Borrow? The Long Decade of Neglect in Nomadic Education   165
10. Bending and Borrowing in Mongolia, and Beyond   185

Notes   205
References   217
Index   241

List of Tables, Figures, Maps, and Photographs

Tables

1.1	Comparative Case Study Analyses and the Study of Globalization	3
1.2	Empirical Studies Used in the Book	19
2.1	The Four Eras of Educational Import Prior to 1990: Overview	26
6.1	Core Educational Patterns in the United States, Japan, and Russia	128
7.1	Outcomes Contract, Example Töv Province	142
9.1	Boarding School Enrollment, Period 1990–2002—Figures for Mongolia	169
9.2	Official Statistics on Dropouts, School Year 1991/92 to 2004/05	178
9.3	Comparative Figures on Dropout Rates—School Year 2003/04	181

Figures

1.1	The Epidemiological Model of Global Dissemination	10
5.1	Percentage of GDP Allocated to Education	91
6.1	The Hierarchical Setting in the Mongolian Classroom	115
6.2	Percentage of Overall Monitoring Time Spent on Various Tasks, by School Level	116

Maps

5.1	Map of Mongolia's Region as Defined by the UN System	90
5.2	Map of the Former Socialist Region from which Mongolia is Excluded	90

Photographs

3.1	Bypassing Capitalism	52
6.1	Lenin *Bagsh* with Teachers from a Rural District-School, Övörkhangai	120

Acknowledgments

For years, each of us was conducting research on educational reform in Mongolia from separate corners of the world. Already accustomed to being regarded as eccentrics within comparative education, we met in Berlin in 2002 and realized there was at least one other person fascinated with educational developments in Mongolia. Four respected colleagues at Humboldt University, Berlin, made this fortuitous meeting happen: Jürgen Schriewer, Jürgen Henze, and Martine Tarrieux in the Department of Comparative Education, and Uta Schöne, Director of Mongolian Studies in the Institute for Asian and African Studies. Without them, the chapters of this book would never have come together under the same cover.

Given the distance from New York and Berlin to Ulanbaatar, it can be difficult to maintain close personal and professional ties with Mongolia. While several Mongolian and international institutions helped us bridge that distance, we would like to highlight two, in particular, that both literally and figuratively kept us going over the past years: the Mongolian Foundation for Open Society (MFOS) and the Achlal Project in Ulaanbaatar. The initial contact with MFOS was established by Liz Lorant, Open Society Institute New York, who, in 1998, managed to convince Mongolian colleagues—inundated with short-term consultants from abroad—to take a closer look at these international consultants/researchers who seemed committed to Mongolia for the long run. A good long run it has been, thanks to Liz. In 2004, the MFOS was dissolved and replaced with several Mongolian nongovernmental organizations. Two of them remained as our home bases in Mongolia: the Mongolian Education Alliance (MEA), directed by Natsagdorj Enkhtuya, and the Open Society Forum, directed by Perenlei Erdenejargal. Two projects of the MFOS/MEA, School 2001 and Teacher 2005, profoundly shaped our assessment of how "traveling reforms" are encountered in Mongolian schools and in teacher education. Additionally, the Achlal Project, directed by Davaanyam Azzayaa, helped us ground our work in the experiences of families living in the poverty-stricken shantytowns surrounding Ulaanbaatar. The Achlal Project offers schooling to dropouts in the Bayankhoshuu city-district, and provides support for disadvantaged families in Ulaanbaatar. Without much publicity, the project has been run for ten years on a low budget while making a great impact on the education, health, and self-esteem of several hundred children and families. Only in 2004 did Achlal receive international funding from the Global Fund for Children, based in Washington, DC. Azzayaa, Enkhtuya, and Erdenejargal are not only colleagues, but also close friends. The work that they have done, and continue to do, is the inspiration for this book.

The Mongolian National University of Education and the Mongolian National University became important venues for collaborating with the best education experts in the country. We see ourselves as students (*shav'*) of many visionary Mongolian experts in education and history including Ookhnoi Batsaikhan, Tsedenbal Batsuur', Nadmid Begz, Sharav Choimaa, Tümenbayar Dashtseden, Süren Davaa, Yadam Ganbaatar, Amgaabazar Gerelmaa, Badrakh Jadamba, Onorkhan Kuliyash, Ochirjav Myagmar, Dondovsambuu Purevdorj, Byambajav Purev-Ochir, Balchinbazar Sumy"aasüren, Sengedorj Tümendelger, and Dagiisüren Tümendemberel. Without these teachers, this book would have been two hundred pages shorter. We are also grateful to Badrakh Jadamba and Johnny Baltzersen who in 2004 gave us access to the 40 partner schools of the DANIDA-funded Rural School Development Project. Over the course of writing this book, we often recalled the visits to those schools, especially the long trips by jeep that we took to reach them. During these trips, we found ourselves debating the future of educational reforms in Mongolia. Thankfully, Gerelmaa, Jadamba, and Johnny always had an opinion, making our discussion come alive and taking the chill out of the evening air in remote villages. At a time when other international donors abandoned rural schools, they mobilized funds to keep the schools networked, and added legitimacy to the tireless and valuable work being done in the rural areas of Mongolia.

Having relied for years on the insights of Mongolian colleagues, it was important to us to give something back by sharing our research with those working in Mongolia. The Open Society Institute (Budapest) and the Open Society Forum (Ulaanbaatar) supported our plan without hesitation. Thanks to the directors of these two nongovernmental organizations, Perenlei Erdenejargal and Katalin Koncz, this book will be translated into Mongolian, reaching colleagues and decision makers in Mongolian education who can really make a difference.

The faculty and doctoral students in our two home institutions have been very supportive of this collaborative research project. We would like to acknowledge, in particular, the support of Ingeborg Baldauf, Director of the Institute of Asian and African Studies (Humboldt University, Berlin), and colleagues in the Department of International and Transcultural Studies (Teachers College, Columbia University, New York). The Thursday doctoral seminar group at Teachers College, Columbia University, showed such great enthusiasm for the project that they ended up hearing more about policy borrowing in Mongolia than they ever asked for.

In the final stages of the book, we relied on several individuals in the United States who provided us critical feedback. Nicole Angotti, Eric Johnson, Manu H. Steiner, and especially William deJong-Lambert, made numerous suggestions as to how better illustrate our accounts and sharpen our interpretations.

Acknowledgments

Even though many scholars don't like to acknowledge their weaknesses, they tend to know their strengths well. In this book, one of us is a specialist in international education policy studies, and the other in Mongolian studies. We hope that, by virtue of being a team of two authors, we have eliminated many of the misconceptions about educational reforms and development in Mongolia. If we haven't, we always have each other to blame.

At Palgrave Macmillan, Amanda Johnson has accompanied us throughout. From her initial support for the book proposal, to her skillful review of the manuscript, she has shared our conviction that there is something to be learned by looking beyond traditional horizons and studying traveling reforms in Mongolia.

Gita Steiner-Khamsi
Teachers College, Columbia University,
New York

Ines Stolpe
Humboldt University,
Berlin

1

Going Global: Studying Late Adopters of Traveling Reforms

This book represents an attempt at understanding why so many educational reforms in Mongolia have been tailored after reforms from elsewhere. Globalization does capture, in a very broad sense, what has been occurring in Mongolia. This trendy characterization suffers, however, from many shortcomings. Among other deficiencies, it is devoid of agency, process, and rationale. Who drives the import of educational reforms? How does policy transfer to Mongolia from elsewhere occur? Why do certain global reforms resonate in Mongolia? Why not others? These are the kinds of questions that help us investigate why educational reforms in different parts of the world, including in Mongolia, are becoming "strikingly similar" (Samoff 1999: 249). While the script of this book may tell a story of globalization and Mongolian education, the fine print traces complex traveling reforms that landed, in some cases with considerable delay, in Mongolia.

The Case: Why Mongolia?

Intuitively, one would not expect large-scale policy import in Mongolia. One-third of the population consists of nomadic pastoralists and another one-third is registered as poor or very poor. Yet most educational reforms in Mongolia are modeled after reforms from high-income countries with sedentary populations. By using Mongolia as a case for studying globalization in education, we adopt a somewhat counterintuitive methodological approach in that we select an educational system that seems, at first glance, least likely to engage in policy import. However, despite all expectations, policy makers in Mongolia actively and enthusiastically engage in policy borrowing. Why Mongolia? The question is twofold: why did *we* select Mongolia as an intriguing case study of globalization in education, and why is educational import so common *in* Mongolia?

A Strong Case for Studying Globalization

With 2.4 million inhabitants in a territory half the size of India, and a population density of 1.5 people per square kilometer, Mongolia is one of the least densely populated countries in the world. Mobility, sparse population, hostile environmental conditions, seasonal migration, and the remoteness of herder families, traditionally constituted the main challenges for securing universal access to education. With the collapse of the Soviet Union and the socialist Council for Mutual Economic Assistance (CMEA), and the subsequent decrease of external financial assistance in the 1990s, maintaining universal access has become an issue. The gross enrollment ratio in basic education (grades 1–8) is still high when compared to other low-income countries, but it has decreased dramatically in the postsocialist era. In 1990, the ratio was 99 percent as opposed to 89 percent 10 years later. Two-thirds of the children that are not attending school, or drop out of school, are boys. Mongolia is one of the few countries in the world where the educational attainment of males is significantly lower than that of females and where the next generation is less educated than that of the parents' generation. The gender and generation gaps are not the only features that set education in Mongolia apart from other countries. Coming to grips with nomadic education, for example, is another major and unique challenge. There are abundant additional distinct features in Mongolian education—some of which are related to the postsocialist, nomadic, and Central Asian education space it inhabits—which urge us to ask why reform strategies from other countries were seen as a panacea for resolving local challenges in the education sector.

A central question in globalization studies is whether educational systems are abandoning their distinct cultural conceptions of "good education" or "effective schools," and are gradually converging toward an international model of education. One of the explanations most frequently given for the international convergence of educational systems is the following: Once the barriers for global trade are eliminated, anything can be imported and exported, including educational reforms. Since the trajectory of that trade tends to be unidirectional—transporting educational reforms from high-income to low-income countries, and rarely the other way around—educational systems in different parts of the world are increasingly becoming similar. It appears easier to describe the features of such an international model of education than to actually name it. Attempts have been made at various times to find an appropriate label for the convergence process. The classic sociological explanation has been modernization, followed by Westernization, neocolonization, Americanization, and McDonaldization. For the past decade, authors have equated the international model of education with the neoliberal model of educational reform (e.g., Henig 1994).

The problem with labels is not the associations they invoke, but the problem lies with the worldview or grand theories to which one has to subscribe to believe them. Each label reflects a particular view of dependence, hegemony, and exploitation, and excludes other perspectives that are necessary to explain convergence in a particular context. Regardless of terminological and theoretical disputes, the fact remains that the idea of *education sans frontière* has been a cause for celebration for some, and a source of anxiety for others. These sentiments are especially pronounced for the

Table 1.1 Comparative Case Study Analyses and the Study of Globalization

	Same Outcomes	Different Outcomes
Most Similar Systems	I Weak case for studying convergence	II Strong case for studying divergence
Most Different Systems	IV Strong case for studying convergence	III Weak case for studying divergence

"strong cases" of convergence, that is, for systems from which one would not expect a convergence toward an international model. Arguably, Mongolia qualifies as such a strong case.

A brief overview of comparative methodology is useful to justify our belief that Mongolia serves as a strong case for examining globalization in education. We present in Table 1.1 the distinction made in comparative case study analyses between systems and outcomes (Berg-Schlosser 2002: 2430; see also Przeworski and Teune 1970), and extend it to the study of globalization in education.

Of course, the terms "strong," "weak," "convergence," and "divergence" are methodological. However, the strength of a case lies, as is discussed in the following section, in its explanatory power.

Quadrant I

Little explanation is necessary as to why educational systems that are similar with regard to their political, economic, and social context move in the same direction of educational reform. For example, the outcomes from the transatlantic exchange of educational reforms between the United States and the United Kingdom during the conservative Reagan/Thatcher era was, although amply documented and meticulously traced, hardly a surprise. After all, the "policy attraction" (Phillips 2004) between the two systems encompassed many areas and was not restricted to choice, privatization, and other market-oriented reforms in education.

Quadrant II

Comparative education researchers often feel compelled to explain the unexpected: Why do educational systems that are economically, politically, and socially similar generate different outcomes? This question is often asked when certain educational systems score lower on international student achievement studies than other systems with comparable standards. For example, the findings of the Third International Mathematics and Science Study (TIMSS) in the mid-1990s generated a huge apparatus of educational studies, and triggered a lively public debate in the U.S. media highlighting the weaknesses of U.S. math and science education as compared to other highly industrialized countries (LeTendre, Akiba, Goesling et al. 2000). Five years later, publication of the league table from the Programme for International Student Assessment (PISA) findings elicited a similar response in Germany. The below-average performance of German secondary school students was not only surprising but was publicly framed as a scandal for the German educational system. Particular attention was given to low performance in reading literacy. Not only did German students score significantly below the average of other OECD educational systems, but the distance between students performing in the top and bottom 5 percent was greater than in all the other 31 participating countries (Baumert, Klieme, Neubrand et al. 2001). Both TIMSS in the United States and PISA in Germany constituted strong cases for investigating divergence with regard to student achievement outcomes.

Quadrant III

The contrastive method of comparison—comparing most different systems that manifest different outcomes—is at the same time the most common and the least informative type of comparison. The Cold War studies of the 1960s, in which researchers from both camps compared their most different systems (United States of America and USSR), as well as the U.S. fascination with the Japanese educational system in the 1980s, were nested in a contrastive research design. Although these studies (over)emphasized differences, their cases were methodologically weak for explaining why the math and science achievements of U.S. students lagged behind those of their counterparts in the USSR and Japan. Left with little explanatory power, researchers resorted to commonsensical reasons by highlighting differences in the larger political, economic, or social context in order to explain different outcomes in the educational system.

Quadrant IV

Our selection of Mongolia as a case for studying globalization is situated here. Mongolia is, methodologically speaking, a case of a "most different system" with "similar outcomes." Finding traces of policy borrowing *even* in Mongolia might be used as strong evidence for an emerging international reform model in education. For example, one could make the point that if vouchers and outcomes-based educational reforms were imported by Mongolia (and they were), then they must have been considered everywhere else too. Obviously globalization has affected educational systems that are similar to each other and therefore prone to "learn from each other." But it has also affected systems, such as Mongolia, that are very different, and thus, at first glance, least likely to benefit from lesson drawing and emulation.

We suggest it is time to pause and think about the possibilities of a case study design that attempts to capture globalization in education: How different is "different" and how similar is "similar," and what are the units of comparison? Qualitative comparative research stands and falls on the selection of cases that are both meaningful for the object of study and commensurable for comparison. Committed to contextual analysis, we feel compelled to present Mongolia as a unique case or a bounded system and to tell the "causal stories" (Tilly 1997: 50) that relate to educational import. At the same time, we are interested in learning from comparison, and thus we render explicit the other cases or systems with which we are comparing educational import in Mongolia.

A Site for Analyzing the Politics and Economics of Borrowing

Mongolia changed political allies in 1990, and the country's move from an "internationalist" (socialist) to an international world-system has had major repercussions for educational import. The postsocialist government has had to learn to speak a new language of reform and has periodically been put under international pressure to act upon it. The new language of market orientation, cost effectiveness, and state deregulation is spoken whenever loans and grants are in sight. These two features—political reorientation and economic dependency—make Mongolia an ideal site for investigating the politics and economics of policy borrowing. Unfortunately, both of these research areas tend to be neglected in globalization studies.

Sociologists at Stanford University, particularly John Meyer and Francisco Ramirez, are regarded as pioneers in globalization studies. As comparative sociologists they have built their argument about globalization on longitudinal studies of educational systems. According to neoinstitutionalist theory, or world culture theory, educational systems have converged not only toward the same "world standards" with regard to the structure, organization, and content of education (Meyer and Ramirez 2000: 120), but also toward the same values of progress and social justice (Boli and Thomas 1999; Chabbott 2003; Ramirez and Meyer 2002). Ramirez writes,

> There are not only more schools and more students (in absolute and relative numbers) than there were at the beginning of the twentieth century, but there are also more common ways of envisioning and interpreting the realities of these institutions. (Ramirez 2003: 247)

The curiosity of scholars in globalization studies with what neoinstitutionalism or world culture theory has to offer is not confined to the question of whether or not reform models in different parts of the world are actually converging toward a singular global model of "modern schooling." They are also interested in whether an adoption of reform models from elsewhere is voluntary or imposed, randomly diffused or systematically disseminated, a complement or a supplement to existing local reforms, and ultimately, good or bad.

Neoinstitutionalist theory has a lot to offer in answering these important questions, but we will restrict ourselves to a critical methodological, comment. Ramirez and Meyer postulate global convergence, but they use countries from a world-system that is one and the same to substantiate their claim. They turn a blind eye to educational systems from other world-systems that are quite different and thereby assume that there is only one world-system. Given the circularity of their argument, the convergence of educational systems within one and the same world-system comes as little surprise. Our methodological critique becomes apparent when we examine the selection of cases on which neoinstitutionalist theory rests. The cases are either countries of the First World, or countries of the Third World colonized by the First World. Methodologically speaking, neoinstitutionalist theory does make an interesting case for international convergence, but its claims rest on weak cases. What is absent from their account is the history of colonization, which would explain some of the similarities between First World and Third World countries, and the history of the Cold War. The Second World, or the other half of mankind (more than 30 postsocialist countries), is conspicuously missing from their list of cases. Until 1990, postsocialist countries inhabited their own, separate world-system. Of course, "progress" and "social justice" had a firm place in socialist value systems, but to be sure, they had a completely different meaning than in capitalist systems.

Anderson-Levitt and her colleagues (Anderson-Levitt 2003) took on the project of scrutinizing the grand claims on which neoinstitutionalist or world culture theory rests, and they did so by juxtaposing it with anthropological notions of culture. As announced in the title of their book, *Local meanings, global schooling*, the authors investigate "local meanings" to visions and pressures of "global schooling," and they find a multiplicity of (local) meanings or outcomes. Their criticism builds on this finding and serves them as evidence for denouncing the homogenizing effects of globalization that neoinstitutionalist theory has asserted. The contributors illustrate that although choice, student-centered learning, outcomes-based education, marketization of schools, and so on went global, they neither replaced already existing models, nor meant the same thing in various cultural contexts. For example, "choice" with regard to the language of instruction, propelled by U.S. missionaries in Tanzania (Stambach 2003) is, for a variety of reasons, a different thing altogether than the "choice" in math instructional methods that factions of PTA associations in California were combating (Rosen 2003). They criticize convergence theories for taking global schooling models at face value without scratching at the surface and examining how they play out differently at the community level. To phrase it more pointedly, convergence theorists seem to have mistaken brand name piracy such as choice, outcomes-based education, student-centered learning, and so on—hijacked from one corner of the world and forcibly moved to another—as heralds of an international convergence of education.

Scrutinizing the claims made by neoinstitutionalist sociologists, Jürgen Schriewer and his coresearchers (Schriewer, Henze, Wichmann et al. 1998; Schriewer and Martinez 2004) remedied the bias in case selection. They acknowledged the existence of several world-systems and examined three of them: Spain, Russia/Soviet Union, and People's Republic of China. What Schriewer's research group at Humboldt University, Berlin, found was diametrically opposed to the Stanford research team. In their longitudinal study of educational research journals (1920s to 1990s), they used the term "references" literally (Schriewer and Martinez 2004). That is, they analyzed the bibliographies published in Spanish, Russian/Soviet, and Chinese journals, and interpreted the type of references made. There is no evidence, they conclude, to suggest that we are increasingly reading the same books and journals in different parts of the world and as a result share the same (international) knowledge on education. What they established instead is a close correspondence between references and political developments in each of the three countries. In other words, whether authors of educational research journals are receptive or hostile toward scholarship from other countries, has to do more with what is going on politically in their own country than with globalization. It is on this question of receptiveness toward internationality that Schriewer and Martinez (2004) make a convincing point. During periods of political isolation, authors either drop their references to scholars from abroad, or else use them in a disparaging manner to distance themselves from foreign influence. They remark that a country's historical and political context (referred to as "socio-logic") is a better predictor of internationality in educational knowledge than globalization (Schriewer and Martinez 2004: 33). In fact, the era of the greatest convergence with regard to educational knowledge was in the 1920s and 1930s, when educational researchers in Spain, Soviet Union, and China were drawn to John Dewey's writing. After that brief period, Dewey was dropped from the reference list in Soviet educational journals and replaced by Krupskaya (Lenin's wife). It is striking that against all expectations of international convergence theorists, educational knowledge in the three countries did not become more internationalized after the mid-1980s, when all three countries opened their ideological boundaries and increased their international cooperation.

For our own study of educational import in Mongolia, we embrace all three of the contributions from globalization scholars just discussed (the neoinstitutionalist focus on long-term trends, the anthropological emphasis on local contexts, and the "socio-logical" receptiveness toward external forces). We also add a fourth that is pertinent to the politics and economics of policy borrowing. The three groups of globalization studies differ, but they complement each other in important ways. Arguably, once we acknowledge the existence of different policy levels, the distinctions made by proponents and opponents of convergence theory become minute, if not obsolete. For example, even though "choice" in education plays out differently in different cultural contexts (Anderson-Levitt's line of argumentation), and resonates for different reasons in different systems (Schriewer's point about the "socio-logic" of selective borrowing), the fact nevertheless remains that "choice" as a concept or a discourse went global (Ramirez and Meyer's conclusion). There is a convergence of educational reforms, but perhaps it is only at the level of brand names, that is, in the language of reform. Once a discourse is transplanted from one context to another and subsequently enacted in practice, it changes meaning.

That said, the points made by all three groups of globalization researchers are well taken, and we should by no means set aside this lively debate among comparative researchers. Each one of them illuminates a different aspect of globalization. For example, it is indeed revealing how a global discourse changes meaning in a local context (Anderson-Levitt), why only specific global discourses resonate locally or "socio-logically" (Schriewer), and how global pressure has been institutionalized in ways to make national decision makers adopt shared global visions of education (Meyer and Ramirez). An inquiry into how global reforms have been indigenized, or "Mongolized," in Mongolia is as intriguing as why only certain traveling reforms have ever made it to Mongolia. We do not attempt to use educational import in Mongolia as a case to recycle what others have already noted. Alternatively, we draw on interpretations provided by the three groups of globalization researchers discussed earlier and offer our own additional proposition. Arguably, our view of globalization is markedly influenced and differentiated by our focus on the politics and economics of policy borrowing.

Most of our own studies so far (e.g., Steiner-Khamsi and Quist 2000; Steiner-Khamsi 2004a) stressed the political reasons for transnational policy borrowing, and we have only started to explore the economic ones (Steiner-Khamsi and Stolpe 2004). In this book we combine, for the first time, both perspectives. There is an aspect of globalization that is often neglected in these studies: convergence in the language of reform as a result of economic necessity, that is, as a result of imposed transnational policy borrowing. This is not an inconsequential point for low-income countries that depend on international grants and loans. In these countries, a portfolio of "best practices," or worse, a complete reform "package," must be imported as a condition to receive funding. As Jones (2004) poignantly notes, international financial institutions are not only in the business of granting loans but also lending ideas. It would be absurd to deny that global pressure in the form of international agreements, a conditionality for receiving external funding, exists in Mongolia. Thus, we do not share Chabbott's celebration of Education for All, and other international agreements, as a herald of a new era, in which all governments voluntarily adopt the same international visions for education (Chabbott 2003). Such an interpretation does not sufficiently take into account the economics of transnational policy borrowing.

This is not to suggest that global pressure is a static entity that is forced upon passive, local victims. Any encounter involves at least two actors, and agency needs to be acknowledged for both sides, that is, for international donors as well as for local recipients. How local forces encounter global pressure, or what makes them adopt, resist, or undermine external pressure on domestic educational reform, is a terrain that deserves far more exploration. Furthermore, agency-oriented studies need to recognize a multiplicity of agencies both among international donors and local recipients. Neither are all international donors in cahoots, pushing the same development agenda, nor are all Mongolians *unisono* either for or against policy import. Similarly, there is great variation at the local level between what politicians pronounce as a fundamental reform, what government officials subsequently legislate in policy documents, and what practitioners eventually implement at the school level. This distinction between "policy talk," "policy action," and "policy implementation" (Cuban 1998) should accompany the reader throughout this book, as it serves as an analytical tool for conducting agency-oriented policy studies.

As mentioned before, convergence often occurs exclusively at the level of policy talk, in some instances also at the level of policy action, but rarely at the level of implementation. This leads us to suggest that more attention should be drawn to the politics of educational borrowing. In Mongolia, we found all kinds of local encounters depending on the type of reform: adoption or voluntary borrowing, open resistance to externally imposed reform, and more subtle ways of undermining reform packages transplanted by international organizations. In this book we introduce our own interpretive framework for studying these local encounters with global forces. It is a framework that takes into serious consideration the politics and economics of policy borrowing.

An Example of Secondhand Borrowing and Late Adoption

The idea of comparing the rapid global dissemination of school reform models to epidemics is not new (see Levin 1998). But it is novel to systematically apply an epidemiological model to explain why, from a plethora of school reforms, only a few appear in different corners of the world. In addition, lately the reforms that have resurfaced in different parts of the world, including in Mongolia, have been neoliberal ones. Thus we ask: What accounts for a contagion, that is, which features of a reform enhance its exportability, and what are the preconditions for transnational policy attraction or import? The analyses of social networks, and in particular Small World research (Watts 2003), as well as earlier studies mapping the diffusion of innovation process (Rogers 1995; see also Gladwell 2002), have much to offer in the way of understanding such reform epidemics. The epidemiological model assumes a lazy S-curve, depicted in Figure 1.1.

Prior to the take-off point, only a few educational systems are "infected" by a particular reform epidemic. At that stage, the early adopters of a reform make explicit references to lessons learned from abroad, in particular, from the reform that they are emulating. A good case in point is the transatlantic transfer of "choice" between the educational systems in the United States, England, and Wales in the early 1990s. A myriad of studies were produced examining how the choice reform functioned in other systems (e.g., Chubb

Figure 1.1 The Epidemiological Model of Global Dissemination

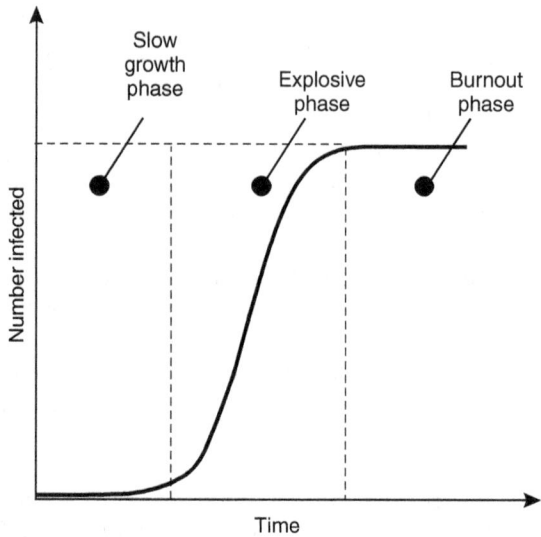

Source: Watts (2003: 172).

and Moe 1992), that is, how the early adopters of choice made explicit references to positive experiences in similar systems. Needless to say, this lesson drawing, or externalization, has a salutary effect and helps to certify a reform that otherwise would have been contested (Steiner-Khamsi 2004b). During the phase of explosive growth, however, more systems adopt a reform, and the traces of transnational policy borrowing disappear. Once a critical mass of such late adopters have borrowed a particular reform, the geographic and cultural origin of the reform vanishes, lowering the threshold for the decontextualized and de-territorialized version to spread rapidly to remaining educational systems. Global dissemination occurs at this stage. An epidemic ends, or the phase of global dissemination ceases, when most educational systems have already selectively borrowed bits and pieces of the reform and thereby generated immunity from other aspects of the reform.

Early adopters are those educational systems that emulate reform from elsewhere during the slow-growth phase, whereas late adopters join a reform movement after a substantial number of systems have already imported the reform. Very often they do not borrow from the original(s), because the educational systems that initiated the reform have moved on to implementing new reforms, but borrow secondhand from other late adopters. Figuratively and literally, Mongolia must be viewed as an intriguing case of secondhand borrowing. Nevertheless, Mongolia, and the rest of the postsocialist world, are late but significant adopters of global reforms. With such a huge mass of educational systems joining the chorus of neoliberal reforms, the few remaining systems that until recently resisted such reforms have little chance of staying immune to them.

The Area

It is easier to situate Mongolia geographically than it is to place it politically, economically, and culturally. Bordering Russia and the People's Republic of China, Mongolia is a Central Asian state. Although located in the center of Asia, Central Asia is often neglected in historical accounts. André Gunder Frank (1992), renowned author of world-systems theory, ends his book *The centrality of Central Asia* with an appeal to recognize the importance of this region for studying long-term changes in world-systems:

> May this help convert Central Asia from the sort of dark hole in the middle that it was, to a real black hole whose gravitational attraction can soon engulf the outside and outsiders. (Frank 1992: 52)

There were periods in history when Mongolia and other Central Asian countries saw themselves as part of the same world, such as in the twelfth and thirteenth centuries, under the Mongol Empire, and in the twentieth century, as "fraternalist" socialist states. These are but two of the eras during which the countries of Central Asia inhabited the same geopolitical space. There are, however, vast differences between Mongolia and other Central Asian countries. As Christopher Atwood (2004) astutely points out, there are three groups missing in Mongolia: Russians, Muslims, and Turks. Instead, what Mongolia has to offer are Chinese, Buddhists, and Mongols. Atwood presents his poignant summary based on a comparison with the former Soviet republics of Central Asia, notably with Kazakhstan, Kyrgyzstan, Tajikistan, Turkmenistan, and Uzbekistan. Furthermore, when he makes reference to the presence of Chinese as a signpost of Mongolia, he has Inner Mongolia rather than (Outer) Mongolia in mind. Nevertheless, Atwood's observations serve as a valuable framework to situate Mongolia within its region.

Mongolia, an independent country with its capital Ulaanbaatar, was—in contrast to the five aforementioned Central Asian countries—never a Soviet republic, and therefore not submitted to a Russification or assimilation policy. Nonetheless, with the revolution of 1921, the residents of the Mongolian People's Republic grew up learning that the Soviets were an "older brother" *(akh)* who protected them in the early revolutionary days against Chinese invasion and colonization, as well as "Japanese aggressors" during World War II. After the war the Soviet Union was credited for actively supporting their Declaration of Independence from China. Every child in Mongolia learned that the Soviets helped the country to safeguard its autonomy and remain an independent socialist state. It would be inaccurate to say that the Mongolian People's Republic functioned as the Sixteenth Soviet Republic, but it would be equally wrong to deny the economic, political, cultural, and military dependence of Mongolia on the Soviet Union and other socialist countries. To continue with the list of differences between Mongolia and other Central Asian states, Muslims in Mongolia constitute only 4.3 percent of the total population (Mongolia National Statistical Office 2001) and are mostly

Kazakhs who settled in the western provinces of the country. Finally, Turkish influence in Mongolia has been minor compared to other Central Asian states. Besides the Kazakhs, there is only one other small Turkic-speaking minority, Tuvineans, who live in the northwestern part of the country.

Consider what Mongolia, according to Atwood (2004), has to offer. Atwood's reference to the Chinese calls for an explanation of "Greater Mongolia" that encompasses Mongolia, several regions in the People's Republic of China, including Inner Mongolia, and the regions in Russia inhabited by Buriat Mongols. In the People's Republic of China, Mongols live in the Autonomous Region of Inner Mongolia, and furthermore they constitute a large minority in Manchuria (especially in the Barga and Daur region) as well as in the Xinjiang province. From a pan-Mongolian perspective, the region inhabited by the Buriat Mongols in Russia is also considered part of Greater Mongolia. In all these regions, the Mongols have been exposed to, and resented, massive Sinicization and Russification policies. Not surprisingly, these assimilation policies gave birth to some of the most ardent champions of pan-Mongolism (Bulag 2002). Atwood's reference to the overwhelming presence of the Chinese thus applies only to Inner Mongolia, and not to (Outer) Mongolia or to the other regions where Mongols live. The ethnic composition in Inner Mongolia changed in the early twentieth century with the large influx and colonization by Chinese farmers and merchants (Kotkin and Elleman 1999), and since then the Chinese population has represented an ethnic majority. Similarly, the ethnic majority in the Buriat regions of Russia are nowadays Russians, and not Mongols.

In this book we focus exclusively on Mongolia. Beginning with the period of Manchu rule (1691–1911) and lasting until 1924, the territory was known as "Outer Mongolia" or "Northern Mongolia," and covered the current territory of Mongolia as well the adjacent region of Uriankhai, nowadays referred to as Tuva. In 1911, at a time when the Manchu Empire started to weaken and dissemble, the opposition in Outer Mongolia grasped the opportunity to declare its independence. This period of independence was relatively short-lived: Already in 1918–19 Chinese troops invaded Outer Mongolia, and were only expelled in 1921 with the support of Russian troops. After entering the Soviet sphere of influence, the government adopted a new name, Mongolian People's Republic, which was used from 1924 until 1990.

A comment is in order on the next feature distinguishing Mongolia from other Central Asian countries. The dominant religion is a Tibetan, Lamaist version of Buddhism. As a result, the Mongolians have maintained strong ties with Tibet, and in contrast to Mongolian–Chinese relations, their relationship has never suffered from a history of war and colonization. Whereas Tibet constituted the religious bond, Russia was Mongolia's most important political ally from 1921 to 1990. This political orientation toward Moscow was also echoed in cultural domains because it implied an emulation of values and beliefs associated with (socialist) Europe. To date, the orientation toward Europe is most visible in Ulaanbaatar where comments on the "Un-Asianness" of the city are often heard.

The final feature that Mongolia has to offer is an epitomization of the nomadic lifestyle. The term "Mongol" provokes, as Uradyn Bulag has discussed (2002), all kinds of associations, ranging from romanticizing ("naturalistic people") to masculinizing ("the untamable") to de-civilizing ("barbarian") notions of nomadic life. In Kazakhstan and Kyrgyzstan (Fratkin 1997; Rottier 2003), sedentarization was the primary native reaction to nineteenth-century Russian settlement, ensuring that the indigenous population was not stripped of their own land. Sedentarization was later enforced by Soviet modernization and industrialization. Meanwhile in Mongolia, nomadic pastoralism was preserved undiminished as a result of Soviet-style modernization. In Mongolia, the collectivization of livestock and agricultural land went hand in hand with building a rural infrastructure to ensure that nomadic herders and workers on agricultural collectives had an equitable standard of living and remained in rural areas. After all, the national economy depended on this rural network of animal husbandry and agricultural collectives. Carolyn Humphrey and David Sneath (1999: 179) coined the term "urbanism" to describe the elaborate infrastructure and urban culture that was transferred into rural areas. In stark contrast to this type of "urbanism" in the rural areas, the post-1990s has entailed "urbanization." With the dissolution of the collectives, the financial means and the political will to preserve an urban-like infrastructure were lost, and an unprecedented internal migration process from rural to semi-urban and urban areas was set in motion.

Until 1990 Mongolia oriented itself politically toward policies emanating from Moscow, and nowadays the country directs its attention to Brussels, Canberra, Manila, Tokyo, New York, Washington, or wherever else the headquarters of international donors are based. As mentioned in the previous section, this reorientation in "space" (Nóvoa and Lawn 2002) is crucial for our study on the politics and economics of policy borrowing in Mongolia. In particular, the shift from internationalist (socialist) to international external assistance has had more of an impact on educational reforms in Mongolia than any domestic developments, and it is a recurring theme underpinning our analyses.

Though culturally an outlier within Central Asia, Mongolia's educational reforms both before 1990 and after have been entirely in line with what other socialist or postsocialist countries have been experiencing. Since the next chapter reflects on the uniqueness of educational development in Mongolia, we take the opportunity in this introductory chapter to highlight some of the similarities with other postsocialist countries.

An anecdote might be an illustrative prelude to the more systematic analysis to come. In July 2004, we attended a memorable meeting of nongovernmental organizations, mostly staff members of policy centers, representing over 15 postsocialist countries, funded by the Open Society Institute (Soros Foundation). The meeting was held in Tblisi, Georgia, and the Georgian Deputy Minister of Education delivered the keynote address. She listed all the accomplishments of her government over the past decade: extension of schooling from 10 to 12 years leading to a school entrance age of 7 rather than 8 years, reduction of the number of subjects in the school curriculum, introduction of new subjects (English and computer literacy), student-centered learning, electives in upper secondary schools, standardized student assessment, reorganization of schools by either closing down small schools or merging them with well-equipped large schools, decentralization of educational finance and governance, liberalized regulations for textbook publishing, and private sector involvement in higher education.

The audience nodded. What was unfolding in front of their eyes was a "postsocialist reform package" (Silova and Steiner-Khamsi 2005), traveling across the entire region of Central, Eastern, and Southeastern Europe, Caucasus, Central Asia, and Mongolia, not to mention their own country as well. The "traveling policies" (Lindblad and Popkewitz 2004) had the same objective—the transformation of the previous Soviet system of education into an international model of education—designed by international financial institutions and organizations. This model was imposed in a few cases, but for the most part it was, in subtle ways, voluntarily borrowed for fear of "falling behind" internationally.

At the Tbilisi meeting, the participants noticed that there seemed to be a "canon of technical assistance" (Stolpe 2003: 168) that international donors systematically pursue in their respective countries. Their educational sector reviews are not only similar in analysis, but they are also strikingly alike with regard to prescribed reforms (see Samoff 1999). For example, all countries experienced a dramatic reduction in public expenditures on education as a percentage of GDP, a professionalization of education authorities and school directors, and privatization of higher education, to name only a few features of the postsocialist reform package exported to the region. This regional package is supplemented with a few country-specific reforms, such as an emphasis on post-conflict education for war-torn countries, or on gender and education for Muslim countries. The participants in Tbilisi also noticed that several areas were only marginally reformed, such as in-service teacher training, preschools, rural education, or inclusive education targeting students with special needs. This left these fields wide open for nongovernmental organizations.

Our case studies of traveling reforms are not singular instances applicable to the Mongolian context only. As we demonstrate in the concluding chapter, the same package of "best practices" has been transplanted to 33 other postsocialist countries, and prior to the 1990s, to other low-income countries that depended on grants and loans from international donors. Mongolia, as unique as it is culturally, shares a lot in common with other countries of the former socialist space who have been forced to orient themselves toward new political and economic allies. In some cases, the reorientation of the government does not move beyond lip service, and amid much fanfare, a "flag of convenience" (Lynch 1998: 9) is hoisted to secure international funding. Once funding has been obtained, the money is used for other purposes, sometimes for domestically developed reforms. How the government actively and creatively deals with global pressure, and what it signals to international organizations as opposed to what it conveys to its own domestic constituents, is at the heart of our analysis. Deciphering both semantic fields in Mongolian educational reform—the discourse geared toward international allies and the one targeting its own population—promises to illuminate the economic and political reasons for policy import. To do so, one needs to master the languages of the different target audiences, that is, both Mongolian and English.

THE PERSPECTIVE

In their Preface to *Empire*, Michael Hardt and Antonio Negri identify the "modern imperialist geography of the globe" as follows:

> Most significant, the spatial divisions of the three Worlds (First, Second, and Third) have been scrambled so that we continually find the First World in the Third, the Third in the First, and the Second almost nowhere at all. (Hardt and Negri 2000: xiii)

Mongolia was a part of the socialist Second World until 1990 when that world crumbled. Is the Second World really "almost nowhere at all" (Hardt and Negri 2000: xiii), or should we perhaps revise this spatial dimension to include postsocialist countries?

In the years after the fall of the Soviet Empire, social science researchers distinguished between capitalist countries ("old democracies") and countries in transition ("new democracies"), assuming that all former socialist countries would eventually, after a period of transition, become converted to full-fledged capitalism. Starting in the mid-1990s, scholars noted that several practices from the socialist past endure in the present (Barkey and von Hagen 1997). Katherine Verdery (1996: 227) sharply criticized "the teleology of transition" that prescribes what needs to happen for a country to fully embrace a market economy. By the end of the first decade of post-Soviet independence, critics of transitology studies boomed in all disciplines and fields of the social sciences. In comparative education, the linearity of the transition argument came under fire (Cowen 1999), correcting earlier interpretations: If educational developments in the post-1990 period come across as chaotic, it has perhaps more to do with the linearity of our interpretations than with actual reality. Cowen's work is an invitation to reflect on the "rules of chaos" (Cowen 1999).

While, as Christian Giordano and Dobrinka Kostova (2002: 74) cynically note, the "orphans of transitology" have moved on to study "democratic consolidation," a group of social anthropologists have gathered to reflect on what it would entail to apply a postsocialist rather than a transitology perspective (see Hann 2002). For example, renowned scholar of Buriatia and Mongolia, Caroline Humphrey (2002a), addresses the importance of enlarging our analytical framework beyond the "transition" period to gain a more comprehensive understanding of developments in the post-1990 period.

Humphrey states,

> It would be perverse not to recognize the fact that people from East Germany to Mongolia are making political judgments over a time span that includes the socialist past as their reference point, rather than thinking just about the present trajectory to the future. (Humphrey 2002a: 13)

We side with Humphrey's assertion that the legacies from the socialist past do not persist in current infrastructures, or political and administrative-bureaucratic practices alone, but they also function as a cultural lens through which all new innovations and reforms are seen and evaluated.

Finally, there are many commonalities between postsocialist countries that deserve a comparative scrutiny, such as the transfer of the postsocialist educational reform package mentioned earlier in this chapter. For Humphrey (2002a), postsocialism is more than simply a construct, it is a comparative research paradigm that leads us to extend the temporal and spatial dimension of our research. Transitologists often content themselves with reiterating or recycling other authors' descriptions of the events and developments in the post-1990 period, which supposedly have been ruled more by shock and chaos than anything else. Postsocialist studies, on the other hand, apply a longitudinal perspective that encompasses the periods before and after 1990. Spatially, this research paradigm challenges us to draw our attention to other educational systems with the same socialist past, and makes us stop believing that Mongolia's educational development in the post-1990 period has been in any way unique.

The linguistic connection between postsocialism and postcolonialism is not incidental, and there exist two parallels between these two research paradigms. In both cases, the "post" signals a historical period as well as a research angle from which the socialist or colonial map of current practices is uncovered. Verdery (2002) takes the analogy a step further and demands that the history of colonialism should be rewritten to include a more sophisticated analysis of imperial rule; this is an analysis that is detached from black and white conceptions reminiscent of the Cold War. A post–Cold War history of colonialism would explain, for example, why Cuba and Mozambique voluntarily joined the Soviet Empire to demonstrate their independence from other states. Such a project would also enable historians to scrutinize the different imperial strategies. Moscow was not about the accumulation of capital extracted from dependent states; the strategy was much more surreptitious. It aimed at a control of the means of production in dependent states, generating economic interdependence within the Soviet Empire. These are not petty semantic nuances of imperial strategies. Rather, they are essential for making sense of where Mongolia was situated with regard to its larger Second World (-system).

More than two decades after the end of the Cold War, one is surprised to see how deep historical studies on socialist systems are still entrenched in conceptualizations and language inherited from the Cold War. For all their failings, socialist systems valued free and universal access to social services, including education and health care. One would therefore expect that the socialist accomplishment of providing universal access to education for a population that, in Mongolia, is both widely dispersed and mobile would be at least acknowledged. On the contrary, many researchers are soaked in skepticism about anything that functioned well in the former socialist period. For them socialist educational systems were rotten to the core, coercive, and only put in place to systematically and effectively indoctrinate citizens. Of course, not all researchers are as graphic in their clichés as the researcher who asked us the following question in an e-mail:

> I am constructing a theory about how reforms outside of the education sector contributed to the successful development of socialist education. Thus far I have been unable to find anything about how much of a role coercion played in getting families to send their kids to school. It is apparent to me that the socialist government was not above explicitly coercive methods. To your knowledge, were any physical/violent methods used to get kids to enroll in socialist schools? (e-mail communication November 9, 2004)

In reviewing the post-1990 literature on Mongolia, we came across many authors, like the one just cited, who are tainted with conceptions reminiscent of the Cold War. Whereas this group of authors would make us believe that the socialist past is a nuisance that is wished to be completely purged from all current practices, there is an even larger group of authors celebrating Mongols as the last survivors of a species that still engages in shamanism and nomadism. Speaking in a language of redemption, theirs is the project to rescue Mongolian traditions and nomadic lifestyle in light of the rapid urbanization in Mongolia.

The Research

Reflecting on ones' role as a researcher and justifying ones' legitimacy to write on behalf of others, has in the last 20 years or so become justifiably a predicament for researchers who work internationally (see Clifford and Marcus 1986). Each of us has spent at least two years in Mongolia, stretched over a long period of time, and we both bring a different perspective to this collaborative book project.

For Gita Steiner-Khamsi the first encounter with an imported educational reform in Mongolia was in 1998, the year standards-based curriculum reform was introduced. More than two dozen visits followed, mostly as a lead advisor for educational programs of the Mongolian Foundation for Open Society (MFOS),[1] but also twice for the World Bank, and once for the Rural School Development Project of the Danish International Development Assistance program (DANIDA). The two projects for the World Bank entailed analytical work and research, first a Sector Note on access and quality in Mongolian education, and then a Public Expenditure Tracking Survey (PETS) dealing with financial leakages in the Mongolian education sector. The Mongolian National University of Education awarded her an honorary doctorate of education in January 2005. A policy analyst and comparative education researcher who previously worked in other countries, she got involved, and stayed involved, in Mongolian reform projects not purposefully but rather due to a series of lucky coincidences.

Ines Stolpe first visited Mongolia at a critical moment in 1992, when the first sector review was being developed. She completed her Masters degree with a dual major in Mongolian studies and comparative education at Humboldt University in Berlin. In 1997, she enrolled for a semester at the Mongolian National University in Ulaanbaatar, a partner university of Humboldt University, and in 1998 she conducted her first extensive field study on education and nomadism. A Mongolist and comparative education researcher, Stolpe's dissertation deals with rural education in Mongolia and examines, among other things, the transformation of the boarding school system for children of nomadic herders in the 1990s (see chapter 9). In 2004, she was a co-evaluator with Gita Steiner-Khamsi and Amgabaazar Gerelmaa of the large Rural School Development Project that successfully supports 40 schools in remote rural areas and is funded by DANIDA. Having grown up socialist in East Berlin, and having first learned about Mongolia from the sympathetic perspective of another "fraternalist state," she is able to provide a perspective on educational reforms that precedes the postsocialist period of the 1990s. She wrote chapters 2, 3, the first part of chapter 4, and has had a major input in chapter 9.

Two reform projects, both funded by MFOS, have left their deep marks on this book: The project "School 2001" (1998–2001), which included 72 schools nationwide and supported school-based in-service training, peer mentoring, and peer training. The second project, "Teacher 2005" (2002–2005), established school-university partnerships and strengthened, among other things, educational research at the Mongolian National University of Education.

The voucher study, presented in chapter 8, emerged within the context of the School 2001 project, when it suddenly dawned upon the Mongolian project director Perenlei Erdenejargal and the coordinator Natsagdorj Enkhtuya that something was going wrong with the voucher reform: the Ministry of Education reported that the vouchers for in-service training were distributed, but none of the teachers from the 72 partner schools had ever held a voucher in their hands.

The Teacher 2005 project promoted empirical educational research in mixed research teams composed of university lecturers from the Mongolian National University of Education and teachers from selected schools. The research teams were located in four provinces (Bayan-Ölgii, Khovd, Dornod, Arkhangai) and in Ulaanbaatar. While the Teacher 2005 project generated a total of 15 empirical studies, we only include findings from those 4 studies in which we served as principal investigator or researcher, respectively. Our involvement covered all stages of research—design, data collection, analysis, interpretation, and publication. Table 1.2 lists the empirical studies which we frequently refer to in chapters 5–9.

Table 1.2 Empirical Studies Used in the Book

Name of Study	Research Team	Location of Data Collection	Data Base
School-Related Migration	G. Steiner-Khamsi, I. Stolpe, S. Tümendelger	Dornod province-center and 12 rural districts	Interviews in 12 (out of 14) districts and in province center: • Teachers and principals from 19 schools (in 12 districts) • 34 herder family households from 12 districts
Pedagogical Jokes and Classroom Management	G. Steiner-Khamsi, Kh. Myagmar, B. Sum"yaasüren	Bayan-Ölgii, province-center, Ulaanbaatar	Questionnaires, individual interviews, focus group interviews, observation in classrooms: • 124 third year students of preservice teacher education • 26 lecturers of didactics, pedagogy, psychology of preservice teacher education • 20 practicum coordinators and clinic professors at schools
Class Monitor	G. Steiner-Khamsi, O. Kuliyash	Bayan-Ölgii, province-center Ulaanbaatar	Questionnaires, individual interviews, focus group interviews, content analysis of note books of class monitors: • 39 former and current class monitors in Bayan-Ölgii • 48 former and current class monitors in Ulaanbaatar
Teachers as Parents	G. Steiner-Khamsi, D. Tümendemberel, E. Steiner	Övörkhangai province-center and 1 rural district, Arkhangai province-center and 1 rural district	Questionnaires and individual interviews in 2 district schools and 2 province-center schools: • 65 questionnaires • 44 individual interviews

Besides these four studies, explicitly designed as research projects, our examination of educational import has also been informed by numerous program evaluations that we conducted on behalf of MFOS and DANIDA. All research projects and program evaluations were carried out in close cooperation with the Mongolian educational researchers listed in Table 1.2 or mentioned earlier, and entailed meeting with teachers, students, parents, and education authorities across the country.

For traditionally oriented comparative education researchers, there is the question of whether involvement in a project ("technical assistance"), and research, should be separate. This question has troubled us for quite some time, and we came to the conclusion that the divide between *real* researchers (detached) and *applied* researchers (involved) (see Elias 1987), or the distinction between comparative researchers (First World) and development researchers (Third World) has become anachronistic. There is a lot of space between doing large-scale quantitative research (OECD- or IEA-type research) at one extreme and conducting ethnographic studies at the other. One needs to be neither entirely detached nor completely involved. Over the course of our involvement in Mongolia, we took on conflicting roles as advisors, evaluators, and researchers. More often than not, we were participants rather than observers in Mongolian educational reform, and became *involved researchers* who neither were, nor wanted to become, detached from what was going on in Mongolian education reform. This high level of personal involvement in the object of study was both an asset and a liability. We soon became marked as "experts" of Mongolian educational reform.

The expert status had many advantages and one disadvantage. Our status led government officials and staff in international and local NGOs to openly share their concerns and ask us for advice regarding ongoing reforms or projects. Furthermore, the close collaboration with Mongolian researchers and practitioners helped us to identify research questions that satisfied our own academic curiosity and were at the same "hot issues" in the Mongolian policy context. Although the focus of our inquiry was on an institutional analysis of reforms (see Escobar 1995), we depended heavily on Mongolian colleagues to interpret the linguistic nuances used by officials to pronounce and enact an educational reform. Finally, a very practical asset of our role was unrestricted access to information, policy documents, and statistical material that otherwise would have been difficult, if not impossible, to obtain.

It is very important to us that our analyses are shared with, and read by, the educational research community in Mongolia. The findings from the four empirical studies mentioned earlier were translated into Mongolian, published in various research journals, and also compiled as an edited volume (Steiner-Khamsi 2005a). Writing for Mongolian readers, we refrained from theorizing globalization, mostly because these academic debates are unfortunately not fully accessible in the Mongolian language. Rather, our Mongolian publications addressed very concrete policy-relevant issues, such as the overcrowding of schools in urban and semi-urban centers (leading to three shifts rather than the customary two shifts in teaching), problems with retaining students and teachers in remote rural areas, the "statistical eradication" of dropouts from official statistics, or the low salary of civil servants that forces teachers to generate additional income by engaging in all kinds of parallel economic activities, including private tutoring and demanding gifts from parents. A few findings from these studies were discussed in the media and generated a heated public discussion on education development in Mongolia.[2] As a result, we have become known as reformers who study reforms. Besides triggering a public reflection on what went right and what went wrong in educational reforms of the past decade, we were also determined to solicit feedback on our studies and ensure that our interpretations were not offtrack.

Having listed a few advantages of being involved researchers, we also owe the reader a reflection on the disadvantages. Perhaps, the greatest liability of being regarded as an "expert" was the authoritative nature attributed to that status. In Mongolia, expert opinions are beyond contestation or criticism. We found ourselves being treated as "founders of discursivity" (Foucault 1984: 114) who establish the "truth" about educational reform in Mongolia, no matter how wrong the interpretations might have been. At times, we found ourselves being treated as "indigenous foreigners" (Popkewitz 2000: 10), in that we were seen as devoid of any cultural affiliation and used as uncontested external voices to legitimize national reforms in Mongolian education. Therefore we had to periodically reassert our role as researchers who depended on receiving feedback from Mongolian scholars and practitioners. We had to insist on being corrected for all our misunderstandings or biased interpretations that may have resulted from our distant perspectives.

2

EDUCATIONAL IMPORT IN
MONGOLIA: A HISTORICAL PERSPECTIVE

Many recent examinations of globalization begin with an apology from the authors for adding yet another work to an already over-studied subject. Our excuse, however, is that there are few studies on the impact of globalization upon educational reform in Mongolia, and there exist virtually no analyses of educational import in Mongolia from a historical perspective. It is important to recognize that educational import in Mongolia did not begin in the past decade. However, earlier phases are either under-documented or not easily accessible. For a book that attempts to advance research on the politics and economics of policy borrowing they are essential.

We are not alone in insisting that there is nothing new about transnational networks and globalization. Most scholars who conduct historical analyses of these phenomena also acknowledge that more attention must be given to earlier periods of interstate or intercultural transactions. A. G. Hopkin's edited volume *Globalization in world history* (Hopkin 2002) is the product of a trend, summarized by Charles Tilly's assertion that "humanity has globalized repeatedly" (Tilly 2004: 13).

A particular perspective on globalization studies was put forward by world-systems theory. This view must be credited with disaggregating the cluster of countries that engage in transaction and introducing a much-needed discussion on power and the hegemonic relations between them. Whether a country is core, peripheral, or semi-peripheral within each cluster or world-system determines its status, as well as its access to various resources (Chase-Dunn and Hall 1997). Propelling the study of "long-term changes" (Denemark, Friedman, Gills et al. 2000), authors in this field examine shifts in commercial circuits, cultural influence, and political dependencies from one world-system to another. For example, Janet Abu-Lughod (1989) traces world orders from AD 1250 until 1350, and maps a multicentered world consisting of 8 regional commercial circuits. According to this analysis, the "modern world-system" (Wallerstein 1974) has its origins in the sixteenth century, and over time it expanded from Europe to cover the entire globe. After World War I, the system began to bifurcate into separate world-economies, market and planned, and since 1990 has reverted back to a singular, capitalist world-system (Wallerstein 2004). "Modern world-system" is a term often used interchangeably with "world-economy" or "world-empire" to describe a capitalist world order whose survival depends on continuous expansion.

In our work, we emphasize the plural in world-systems theory and apply smaller units of analysis. We envision a cluster of countries whose residents see themselves as members of the same transnational space. Wallerstein's comment on the hyphen in world-system underlines this point:

> Note the hyphen in world-system and its two subcategories, world-economies and world-empires. Putting the hyphen was intended to underline that we are talking not about systems, economies, empires *of the* (whole) world, but about systems, economies, empires *that are* a world (but quite possibly, and indeed usually, not encompassing the entire globe). (Wallerstein 2004: 16)

With this in mind, we distinguish between various world-systems or "spaces" that Mongolia inhabited between the seventeenth and twentieth centuries. Recent educational import is presented separately, and we devote much of the book to reflection on policy borrowing under postsocialist conditions.

Mongolia is an important case study in the world-systems perspective because, depending on the historical period, it has figured as a core, semi-periphery, and periphery in its own world-system. There is a close correspondence between world-systems and educational trade wherein the likelihood of export is greater for core states, and import more common for dependent or peripheral states. No doubt, a more detailed analysis of education under the rule of the Mongol and other non-European empires would be insightful for studies on colonial education that, regrettably, tend to focus exclusively on formal education under European colonial rule. However, such a project would clearly transcend the scope of this book. Instead, we focus on formal education and periods in Mongolian history (seventeenth century until 1990), in which reforms from elsewhere were either imposed or voluntarily borrowed.

Mongolian education traditions developed within a nomadic civilization, lending them a unique cultural profile. Uradyn Bulag has characterized nomadism as "the ultimate cultural symbol defining the core of Mongol identity" (Bulag 2002: 10). The factor "nomadism" is thus a central feature for comparison in our analysis of educational imports. Although all forms of *formal* education in Mongolia were influenced by imports, various autochthonous concepts did exist before educational institutions were established. However, they are not discussed in detail here. Throughout every era of educational import we consider, concepts native to Mongolia may be continuously identified. Once all external influences have been subtracted, these concepts can be characterized as the persistent, culturally specific core of what Gita Steiner-Khamsi (2003) has called a "residuum." In our view, the residuum is not static, but rather a dynamic repertoire of interpretation patterns responsible for the transformation and Mongolization of imported models as they are incorporated.

Historical analysis of the development of schooling in Mongolia suggests that transformational innovations have always been induced by external forces. Developments in the sector of formal education should therefore be traced to constellations in foreign relations, rather than domestic decisions or changes in education policy. External pressures not only served as a catalyst, they have also shaped the content of every educational import to date. These pressures include the interplay of "external powers," aspirations of innovative actors as well as the current *zeitgeist*—"spirit of the times."

In order to account for these three central parameters—politics, agency, and *zeitgeist*—we conduct our analysis from the perspective of distinct eras. This raises the question of historical context—how the influence of earlier imports was manifest alongside indigenous elements. At some point imported innovations became part of Mongolia's cultural heritage and were perceived from within as part of the own tradition. These "invented traditions" (Hobsbawm and Ranger 1983) refer to significant interconnections in the history of Mongolian relations. We assume that receptivity to imports was not, and still is not, solely influenced by the leitmotif of transfer and tension between voluntary import and coercion. Also important was the potential for new concepts to merge with those already in place, altering in the context of nomadism. We also consider the discursive ruptures that resulted from changes in reference society (Schriewer, Henze, Wichmann et al. 1998).

The following Table 2.1 gives a brief overview of the four eras of educational import prior to 1990, delineated according to our criteria for comparison. It is apparent that the first and second eras largely ran parallel to one another. Yet both not only had fundamentally different reference horizons, the implications of each were in many respects diametrically opposed: sacral versus secular, voluntary versus obligatory, prestigious versus un-prestigious. It is also noteworthy that the first era extended into those that followed. Education in monasteries, the first formal education system established in Mongolia, continued into the 1930s along with the establishment of secular schools. It was also, until its eradication, the most widespread form of schooling.

First Era: Enlightenment

The first import of formal education to Mongolia occurred following the successful proselytization of Tibetan Buddhism (Lamaism),[1] starting in the second half of the sixteenth century. There had already been contact with Buddhism before the rise of Chinggis Khan, but during his reign Buddhism played a negligible role in Mongolia. However, some of Chinggis's grandchildren developed an affinity for the religion. Khulagu for example converted, while others such as Möngke were in contact with Tibetan priests, surrounded themselves with representations of Buddhism, and promoted the printing of religious texts.

One of the key events in Mongolia's relationship with Tibet in the thirteenth century was the encounter between the Yuan Dynasty[2] ruler Khubilai, who had already converted to Buddhism, and his spiritual advisor, the

Table 2.1 The Four Eras of Educational Import Prior to 1990: Overview

	Era			
Our Focus	1 Enlightenment Inner and Outer Mongolia	2 Colonialization Inner and Outer Mongolia	3 Nation Building Outer Mongolia	4 Universal Access Mongolian PR
Parameter				
Time	Seventeenth century to 1930s	Inner Mongolia: 1776–1911 Outer Mongolia: 1791–1911	1911–1918/19	1921–1990
Most important reference society	Tibet	Manchuria (Qing Dynasty)	Europe (transmitted via Russia)	Soviet Union
Dominant languages in higher education	Tibetan	Tibetan Manchurian Chinese	Mongolian foreign languages (Russian, Chinese, Japanese, English, among others)	Russian
Receptiveness toward nomadism	positive	positive	positive	positive
New features (selection)	• institutionalized sciences • formal educational institutions • didactic models • a scholarly language • universal access for the entire male population • "vocational training"	• institutionalized training for civil servants • establishment of formal secular schools • more extensive teaching of foreign languages • development of textbooks	• goal: a secular education system for everyone • introduction of modern curricula • the press as a medium for educational "policy talk"	• promoted literacy • coeducation • school system with several grades • preschool • polytechnic education • vocational training • modern sciences • internationalization

Tibetan monk hPags-pa. Together they developed the "two principles" (Mongolian: *khoyor yos*), which envisioned cooperation between the secular (*khaany zasag*) and the sacral (*nomyn zasag*) spheres, and these two principles were set down in the White Chronicle (*Tsagaan Tüükh*) (Bawden 1968; Moses 1977; Baasanjav 1999). This political–religious alliance began, according to Morgan (1986), Rossabi (1988), Baabar (1999), and Shagdar (2000), in 1264 when the Tibetan monk hPags-pa, who had taught high-ranking ministers in the court, was granted the honorary title of "State Preceptor" (*ulsyn bagsh*) by Khubilai Khan. This is important to the history of education in Mongolia for two reasons. First, it established a pattern of relationship in which Mongolian rulers received instruction on religious and spiritual matters from Tibetan priests, accepting them as teachers, and second, it positioned Tibet as the primary reference society.

Contact with Buddhism in the Middle Ages, however, was limited to specific instances. In the following centuries shamanism and popular animistic religions dominated the social practices of the nonaristocratic majority of the population (Lkhagvasüren and Boldbaatar 1999). The subsequent

Buddhist proselytization in the mid-sixteenth century was induced by secular Mongolian elites and influential Tibetan lamas, primarily as a result of power considerations. Mongolian rulers became rivals following the dissolution of the empire, and Buddhism offered an opportunity to gain authority by tapping into the prestigious field of the spiritual. In Tibet, meanwhile, bitter religious feuds raged between the Lamaist schools of the Red Hats (Tibetan: *rNying-ma-pa, Sa-skya-pa,* and *Kar-ma-pa*) and their opponents, the representatives of the Yellow Hats (Tibetan: *dGe-lugs-pa*)[3] reformed by Tsong-kha-pa in the fifteenth century.

Among the Mongols, the Tümed prince Altan Khan (1507–82)[4] from the Ordos region was a key figure, particularly in terms of the earlier mentioned "two principles" (*khoyor yos*) established by the emperor Khubilai and the Lama hPags-pa. In 1578, Altan invited the leader of the Tibetan Yellow Hat school, bSod-nams-rgya-mtso, to his court for spiritual sanction of his ambitions to accede to power. Altan Khan granted the third successor to Tsong-kha-pa the title "Dalai Lama,"[5] which was thus also retroactively granted to his two predecessors. For his part, the new Dalai Lama granted Altan the title of "King of Mind, Very Strong from Heaven." He also appealed to historical references, declaring himself the reincarnation of hPags-pa and Altan Khan the reincarnation of Khubilai. By basing his arguments in theology, Altan legitimized his claims to the throne. Though he died without having realized his ambitions (Bawden 1968; Baabar 1999), Altan can be considered a decisive agent for the spread of Buddhism in Mongolia. He not only erected the first Lamaist monastery on Mongolian territory but he also with the "Code of Altan" in 1569 created laws radically oriented toward the newly introduced religious values, limiting shamanist practices.

There are divergent opinions as to whether or not Lamaist Buddhism was voluntarily adopted in Mongolia. Charles Bawden (1968: 27–28) quotes Chinese sources that reported an enthusiastic reception, while in other passages (Bawden 1968: 32f.) he states that following the period of proselytization there was major repression directed at the representatives of shamanism. Byamba Rinchen (1957: 44), who quotes Tibetan and Mongolian sources, reports that there was massive repression against shamanist practices. Larry Moses (1977: 119) gives examples of influential lamas who carried out unambiguously anti-shamanist campaigns. However, it remains uncontested that Tibetan monks traveled through Mongolia during the sixteenth and seventeenth centuries primarily as wandering priests, focused on converting the aristocratic elite, and it also remains uncontested that this group then took over the task of imposing new teachings in a top-down fashion among the inhabitants of the territories under their rule.

It is surprising that Lamaism not only became widespread in most of Mongolia within two centuries but also soon became a far-reaching institutional authority. Three factors played a decisive role. First, relations between the Tibetans and the Mongols had not been historically damaged by animosity, war, or colonization. Second, a sociocultural affinity existed between Tibet and Mongolia: Tibetan culture was also shaped by the nomadic way of life, to which Buddhist practice accommodated. Third, the relative adaptability to autochthonous belief systems in Mongolia also played a role. In Tibet the Lamaist currents were already hybrid forms that arose from a merging of Mahayana Buddhism and the local animistic Bon religions. G. Lkhagvasüren and J. Boldbaatar (1999) have pointed out that for Mongolia, the process of interreligious merging was not a one-way street. Rather shamanism and Buddhism mutually influenced and transformed one another. Nonetheless, this hybridization of belief systems was limited—as the earlier mentioned bans on shamanist practices show—in cases where either the practices strongly clashed with basic Buddhist values or where shamanism contested the religious and political primacy of Lamaism.

The success of Buddhism fundamentally transformed the Mongolian society in ways that particularly affected education. Buddhist sciences were institutionalized, establishing the first formal education system and creating a national educated elite (Choimaa, Terbish, Bürnee et al. 1999). With the exception of private instruction organized in yurt schools—already in existence in certain areas in earlier centuries (Shagdar 2000)—there were no formal educational facilities in Mongolia until the seventeenth century. It was only once a dense network of monasteries had developed[6] where a new clerical educated elite of lama monks served as teachers that the situation changed. It was customary for almost all Mongolian families to send at least 1 son between the ages of 7 and 10 to a monastic school (Shagdarsüren 1976; Bulag 1998).[7] Study at a monastic school not only offered the novices knowledge but high social prestige as well. The larger monasteries had their own school temples (*datsan*), in which there were three grades (*zindaa*), and in each discipline certain academic and religious titles could be earned (Battogtoch 2002).

With the spread of Buddhist teachings came an influx of translations of canonical literature (Dashdavaa 1999). The largest *datsans* had a comprehensive library with manuscripts and block prints in the Tibetan and Mongolian languages (Montgomery and Montgomery 1999). Sechen Jagchid and Paul Hyer (1979: 227) have noted that studying the Tibetan language was soon more prestigious among Mongols than studying Mongolian. Already in the first era of educational import, learning the language of the exporter was important to the recipients. Although basic instruction took place in Mongolian—as was also the case in all the later eras of educational import—achieving higher honors was predicated from the start upon learning the foreign language of the reference society that served as a model.

Not only new educational content but also new metaphors for education were brought to Mongolia through the medium of an imported language. For example, *gegeerel*, one of the most common words for "education" still used in Mongolian today, literally means "enlightenment." This is fundamentally different however from the European concept of "enlightenment." Etymologically the word is derived from *gegee* (light, shining) and stems from Buddhist terminology: *gegeen* connotes "light/wise/illuminated/enlightened." *Gegeersen khün* originally meant an educated person who strives, through profound knowledge and meditation, for a state of enlightenment according to the ways of the Buddha. The term *gegeerel* later changed in meaning; since the socialist era it has meant both profane education as well as enlightenment in the sense of secularization.

In the monastic schools with Tibetan influences, characteristic teacher–student relationships evolved, shaped by asymmetric normative expectations (Narangoa 1998) and accompanied by certain rituals. The lama assigned as a teacher to a novice student (*shav'*) served as a role model, responsible for education and initiation in the secrets of the Buddhist teachings (*nom*).[8] The didactical approach, that is, the mentoring and teaching model, was tailored after practices in Tibet. Werner Forman and Byamba Rintschen (1967) have sketched the typical program: Everyday following early prayers in the temple, the novice was asked questions about the homework assignment from the previous day. The teacher then read the next assignment from a textbook in Tibetan and explained it in Mongolian. Before commencing the translation exercises, the monastery students had to know prayers and 39 texts in Tibetan by heart. In his study on Buddhism and education in Tibet, Josef Keuffer has shown that "drills" (Keuffer 1991: 50) always constituted the first phase in the learning process before the acquired knowledge was consolidated in discussions. The implementation of monastic didactics found fertile ground in Mongolia and was reinforced by already existing concepts. Rote learning long predated the Buddhist era as all the classic genres of literature were transmitted orally from generation to generation over many centuries (Tserensodnom 2001; Chagdaa 2004).

In addition to its compatibility with autochthonous methods of didactics, two other factors also contributed to the success of the first educational import: (1) universal access to the monastic school for the entire male population, and (2) their integration into the nomadic context. Monastic schools arranged for local room and board in response to the mobility of students' families. This model of school organization was a forerunner of the boarding-school system that developed later, which we consider at greater length in chapter 9. Access and integration were supported by the multileveled system. Similar to the situation in Tibet (Bass 1998), not all students in the monastic schools became scholars. Some of them learned a trade or artistic craft in the monastery's workshops and became makers of profane objects (Taube and Taube 1983). The monasteries thus transmitted not only knowledge in the sciences and languages, they also served as *de facto* "vocational" institutions in old Mongolia.

Second Era: Colonization

Almost parallel to Buddhist proselytizing in the seventeenth century, a political event of fundamental significance occurred that also initiated an educational import: the founding of the Qing Dynasty. The history of the relations between Manchuria[9] and Mongolia had until then mostly consisted of alliances on the basis of marriage or war. The most important event in the cultural exchange between the two entities was Manchuria's adoption of the Mongolian script, starting in the seventeenth century. During this time the Manchurians were becoming stronger both politically and militarily and were pushing into Chinese territory. In 1644 they took Beijing, which subsequently became the center of the Qing Dynasty's power. From then on the Manchurians—despite the fact that they were a Tungusic people (Mongolian: *khamnigan*) and close to the Mongols in ancestry, language, and the nomadic way of life—took it upon themselves to become primary representatives of Chinese civilization in their relations with the border peoples (including the Mongols). The influence of the Manchurians became apparent in Outer Mongolia[10] in 1691, after the Khalkha princes accepted the protection of the Manchu emperor at the conference of *Doloon Nuur*, to receive military support against the western Mongolian Oirad.

The question arises as to which circumstances would justify the notion that Mongolia was colonized by the Manchurians and how this fits into the history of tensions in Mongolia's relations with China. Owen Lattimore (1962) describes the tactics of Manchu politics as double-edged: In relation to China the new rulers represented themselves as heirs to the Chinese cultural tradition, in relation to the Mongols they behaved as if they were in fact closer to them. In this he partly agrees with Gavin Hambly (1991) and Bat-Erdeniin Baabar (1999) who found that the Manchurian rulers in a certain sense took up the political legacy of the Chinese Ming, who had also exploited the rivalries between the Mongolian princes and supported the establishment of theocratic institutions. On the other hand, according to Hambly, the rule of the Qing proved not very significant because Manchurian suzerainty[11] over the Mongols remained largely nominal. This statement is, however, only true for Outer Mongolia that, as Udo Barkmann (1999) maintains, had a special status within the Qing Empire. Practically this meant that the Mongolian princes, despite their status as vassals to the Manchu emperor, still maintained their own laws and tax revenues as an attribute of their sovereignty.

What did things look like in Inner Mongolia? Hambly (1991) takes the position that events experienced by the Mongols as colonization were less the outcome of rule by the Manchurian Qing Dynasty than the activities of Chinese merchants and peasants. According to Hambly, the latter were responsible for heavy debts among the Mongolian population and drove them from the most fertile grazing land with an aggressive settlement policy. He is convinced, however, that the colonization policies of the Chinese were not sanctioned by the Manchurian sovereigns. Bulag (2002: 35) takes a different view, maintaining that "we need to understand the Qing rule as 'colonial.' " His reasoning for this is based first on the successes in integrating faraway areas into the imperial state: Inner and Outer Mongolia were put under strict military control and divided, via a far-reaching administrative reform, into "banners" (*khoshuu*),[12] in which residents were not allowed to cross the borders of the grazing lands. Second, Bulag explicates the complexity of colonial relations in Inner Mongolia. He writes that although the land was not alienable, the Mongolian princes could transfer the rights to its use and thus became "princes-cum-landlords." They lost authority over Chinese tenants, however, as the Manchurian Qing gradually installed a Chinese government, which ruled the Chinese peasants in the Mongolian provinces (Bulag 2002: 108). Bawden (1968: 47), who is right to consider the areas south and north of the Gobi separately, speaks of the "disappearance of . . . independence" in Inner Mongolia, while in Outer Mongolia "independence survived, though even there it was only a qualified independence."

Let us now consider the question of colonization and alliances in relation to the educational import induced by the Qing Dynasty. Taking these two aspects into account we can establish the grounds for the formation of alliances during the phases of educational import that followed. Having referred to the work of historians, historians of education should also be heard so that it is apparent why we describe this era in terms of colonization. The relevant literature (e.g., Sharkhüü 1965; Jigmedsüren/Baljirgarmaa 1966; Shagdarsüren 1976; Erdene-Ochir 1991; Baasanjav 1999; Shagdar 2000) is largely unanimous in claiming that foreign rule by the Manchus had a crippling effect on economic and cultural development of Mongolia. It is interesting that none of the authors refer positively to the fact that Manchurian administration led to the foundation of many secular schools in Mongolian areas. The explanation for this is simple: Ts. Sharkhüü (1965) has shown that the administrative schools established by the Manchurians were not intended for the general public, serving instead as training grounds for scribes (*bicheech*) and civil servants (*tüshmel*) who would work for the Manchurian state. The instructional content was therefore focused primarily on conveying knowledge of writing and law. This was thus clearly about the transfer of a typical colonial tradition of installing natives in lower-level administrative positions in the occupied areas.

Thanks to a collection of original documents compiled by Sharkhüü (1965), events at the level of policy action during the second educational import in Mongolia are well documented for the entire period of the Manchurian occupation. The following provides a brief overview. Already in 1776 the local Mongolian princes received orders to ensure that in every *khoshuu* a few people would learn the Mongolian script. Starting in 1781, the youngest of these students (boys under 17 years of age) were chosen to take a course in Manchurian for at least 3 months. In 1811 it was decreed that for Outer Mongolia a small school would be established at all horse postal stations (*örtöö*) along the traffic and communication routes, serving at least four students. Later, starting in the period between 1851 and 1861, students were trained specifically for office work. In 1898, when Manchurian ministers were themselves increasingly using the Chinese script, it was decreed that a writing school for Chinese would be established in the capital Ikh Khüree[13] where children from different *aimags* (provinces) would be educated (Sharkhüü 1965).

From the perspective of the history of education this suggests that the Manchurian tactic of oscillating between the Chinese and the Mongolian cultural spaces toward the end of the dynasty was conspicuously abandoned in favor of a focus on China. For the increasingly Sinicized and Sinocentric-acting Manchurians, China was without a doubt the political and cultural reference society. But this was not true for the Mongolians, since this second educational import was induced and largely forced upon them by a foreign power. While the Mongolian population willingly sent their children to monastic schools following the Buddhist proselytization, the same cannot be said of the secular schools for scribes and civil servants (*alban surguul'*) set up by the Manchurian administration at almost the same time.

According to Baasanjav (1999), only lower-level civil servants (*khia*) were trained in the educational institutions founded by the occupying power. As a consequence, opportunities for social mobility were extremely limited, and work in the colonial administration was not particularly attractive. Rinchen (1964) has pointed to an interesting form of passive resistance: some students in the writing schools learned how to read, but not to write, so that they could not be forced to work for the Manchurians. What is clear is that the secular schools of the occupiers, and the Manchurian and Chinese languages taught there, did not have nearly the same prestige among the Mongolian population as the monastic schools and the Tibetan language. The latter offered not only the concrete potential for qualifications and social mobility, but also for spiritual promise and a high social status. In this sense nothing changed in the Mongolian orientation toward the self-chosen reference society of Tibet due to the educational import induced by the Manchurian Qing Dynasty.

Nonetheless, the occupiers apparently did not see the monastic schools as rivals to their secular schools. The Manchurian emperors even initiated the founding of monastic schools themselves, indicating that they understood the spiritual and the worldly spheres as complementary entities that posed no threat to one another, as if continuing the Mongolian "two principles" tradition. This does not represent a hybridization in Mongolian culture, but rather the instrumentalization of a political principle that promised to instill "peacefulness" in the once very bellicose Mongolians by propagating Buddhist values.

There was, however, within the Manchurians' own political sphere a force that was perceived as a potential rival: the Russian neighbors to the north. Sharkhüü (1965) has documented that the Russian consulate in Ikh Khüree repeatedly requested permission to found a school for Russian language starting in 1894, and that the requests were repeatedly denied by the Manchurian ministers. In Sharkhüü's estimation, the Manchurians feared Russian influence in Mongolia and tried to hinder any attempts at building relations (see also Barkmann 1999). While the Manchurian occupiers were trying to boycott contact with the northern neighbors, they were at the same time organizing cultural transfer from China, giving preference to their own reference society. According to Jagchid and Hyer (1979), the translation offices of the Manchu administration were not only responsible for official documents, they were also busy translating the Confucian classics into Mongolian to effectively contribute to the spread of Chinese values in Mongolia.

What was the scope of this second educational import? The dualistic teacher–student relationship common to the monastic schools was by and large an exception in the schools for scribes and civil servants. In the latter, a classroom-like structure existed, and given the usually small numbers of students, situations differed from one school to the next. Shagdar (2000: 49–50) reports, in reference to Rinchen, that between 1776 and 1800 only 58 people were trained to be scribes for the Manchurian script. Shagdarsüren (1976: 26) estimates that at the beginning of the twentieth century the number of students in secular schools in Outer Mongolia was 500–600. For the year 1911 more precise figures are available: Shagdar (2000: 67) writes that in the 55 existing *khoshuu* schools, 360 students received instruction, which means only 0.3 percent of the entire population. In contrast, approximately 18,000–20,000 school-aged children went to monastic schools, which represents 25–30 percent of the population (Shagdar 2000: 73).

According to Schöne (1988) and Shagdar (2000), the students of the Manchurian schools for civil servants came from different social classes. The age at which they were accepted into the schools was 15–17 years, and the prerequisites were (1) a bright mind; (2) command of the written Mongolian language; and (3) the willingness of their home *khoshuu* to take over the costs. The first prerequisite refers to the qualifications necessary for entrance. These were most certainly not motivated by egalitarian principles, but instead resulted from the low social prestige of the schools for civil servants. The second prerequisite points to another, more pragmatic reason why the Manchurian occupiers did not see the monastic schools as competitors: along with the simpler scribe schools, they seem often to have supplied basic literacy to those who would later become civil servants. The third prerequisite refers to the financial model used to pay for this type of education. In true colonial style, the student's received financial support from the local community (Shagdar 2000).

The new positive features in the second educational import era include— in addition to more extensive language instruction—the impulse for increased translation activity and textbook development. Because the schools for scribes and civil servants could not rely on imported textbooks for instruction, Mongolian teachers and writers developed a selection of primers and story books. In the nineteenth century in particular, many textbooks for history, geography, and translation were developed, which were used well into the next century (Baasanjav 1999).

The Manchurian response to nomadism was, as mentioned earlier, to divide Mongolia into territorial units (*khoshuu*) where the inhabitants were assigned. This gave the Manchurians administrative and military control over their mobility. But as Barkmann (2000) has pointed out, it was not in the interest of the Manchurian imperial court to transform nomadism into sedentarism, given that nomadism is necessary for successful animal husbandry. Schools for scribes and civil servants were established along existing fixed points, such as horse postal stations along courier routes, which also served as administrative centers. Students from far away areas received room and board near the schools (Sharkhüü 1965).

In conclusion, it can be said that the institutionalization of secular education—despite being a product of colonialism—brought with it a considerable emancipatory potential as an unintended side effect. More than a few graduates of the scribe and civil servant schools were later able to use the knowledge and skills they acquired in their struggle to attain and defend the autonomy of Outer Mongolia. Making access to education not dependent upon the social class of the students was something that carried over into the two following eras of educational import, during which the inclusion of the female population became a deliberate goal and served as a sign of modernization.

Third Era: Nation Building

When the Qing Dynasty fell in 1911, the educated elite in Outer Mongolia used the chaos in China as an opportunity to strive toward autonomy. The short-term goal was the creation of their own state structures, the long-term goal was the realization of the idea of pan-Mongolism—the unification of all Mongols, partially living under Chinese and Russian rule. Initially the planned declaration of autonomy in Outer Mongolia was linked to the necessity of forming an independent government. The eighth *Javzandamba Khutagt* was enthroned as the theocratic leader of the new state. As the highest-ranking incarnation of the "living Buddhas" in Mongolia, he had the greatest political authority and was unchallenged by any rivals to the throne of the Great Khan. Tibetan by birth, he was a spiritual ruler and worldly leader, with the title *Bogd Gegeen* or *Öndör Gegeen* ("Lofty Brilliance"). In this capacity he created an autonomous government based on a Western model with five ministers. The form of government was imported from the West, and the international term "autonomy" (*avtonomi*) was used to gain (Baabar recognition from the outside world 1996 and 1999).

When independence was declared in December 1911, the aim was to make a radical break with China. In 1912, a Russian–Mongolian treaty was signed for the primary purpose of securing Outer Mongolia against Chinese troops and colonization by Chinese settlers. In the context of this threatening scenario relations between Mongolia and Tibet, which had in the meantime developed into a tradition of friendship, were strengthened by a treaty of mutual assistance in 1913. The treaty had three main points: (1) mutual state recognition; (2) acknowledging the *Bogd Gegeen* as the leader of Mongolia and the Dalai Lama as the leader of Tibet; and (3) a statement calling for the further promotion of Buddhism. After tough diplomatic wrangling a treaty was also signed between China and Russia in 1913, affirming the political status of Outer Mongolia for the time being. China recognized its domestic autonomy, and Russia recognized that Outer Mongolia was under the suzerainty of China and could not conduct its foreign affairs autonomously. Though the declaration was made without Mongolian approval, it established the country's independence and created de facto autonomy long before the de jure autonomy provided by a referendum following World War II.

While the Great Powers were negotiating the fate of Mongolia, the Mongolians were preparing to enter the international arena as a modern state. Their declaration of independence was sent to France, Great Britain, Germany, the United States, Japan, Denmark, Holland, Belgium, and Austria-Hungary. However, there was almost no response, and the political actors did not anticipate that, as Baabar (1999: 103) put it in relation to the very late entry of Mongolia to the United Nations, ". . . it would take thirty-five to seventy-five years for these powerful countries to recognize Mongolia's independence."

What were the implications of the nation building process for the education sector? First, the newly independent state desperately required qualified, native experts to perform civic duties and secure autonomy. This led to the founding of a central institution (*Mongol ornyg shinjin üzekh tukhain khereg shiitgekh khoroo*), specifically responsible for questions of modernization (including in the sector of education). It is significant that the secularization of the education sector began under the aegis of a theocracy whose leader was a living Buddha, Tibetan by birth. Although—as described earlier—relations with Tibet were strengthened politically and religiously after independence, Tibet no longer had priority as a reference society for Outer Mongolia in terms of education. This is because secular education was in demand, and culture affinity less important during the nation building process. These factors triggered the active reception of foreign models, primarily from Europe—via Russia—which became the new reference horizon.

The educational policy strategies of the autonomous government, based on the Western model, were aimed at establishing a state school system. Although Mongolian autonomy was only to last for 8 years, 111 state decrees were issued pertaining to education.[14] In 1912 the first state school was opened in the capital Niislel Khüree.[15] The location of the school, in the foreign ministry building, was significant as this ministry became the institution primarily responsible for questions regarding education. In the same year the foreign ministry addressed an official letter to the administrators of the capital city as well as to the four aimags, in which foreign orientation was justified with the argument that contact with the world outside could only be established by fostering the qualifications of native Mongolians. This meant foreign languages were given educational priority. As implied in the letter, Buriat intellectuals were considered to be the new experts who were to tackle the myriad tasks of the nascent state (Jigmedsüren and Baljirgarmaa 1966). With their European education and as representatives of a Mongolian group that had traditionally lived on Russian land, the Buriats were predestined for the role of political advisors and could serve as translators and mediators. Once Mongolia achieved autonomy, the classic pattern of employing Tibetan lamas as advisors on political questions began to alter. A new practice emerged that would become the standard for the fourth era of educational import: the Buriats were the first "teachers from Russia."

An exemplary figure was the scholar Jamsrany Tseveen. He was brought to Mongolia to serve as the official in charge of cultural issues in the foreign ministry after he had written to the government concerning the necessity of a multilevel state school system (Rinchen 1964; Idshinnorov 1997). Another Buriat, Erdene-Batukhan, also took on a leading role as an advisor for the establishment of state schools during the period of autonomy. Under their direction, foreign ministry officials looked to Russian models for education policy, developed with reference to Europe, and thus considered "modern." According to S. Jigmedsüren and B. Baljirgarmaa (1966), they outlined the basic guidelines for the statute and curricula according to the paradigm of Russian elementary, secondary, and even postsecondary education. In 1913 the first Russian school was founded in the capital and a second one in 1917 in the western Mongolian city of Ulaangom (Shagdar 2000: 79).

In the rural areas, in addition to the existing 60 scribe schools, at least 49 state primary schools were established by 1917. They were largely housed in yurts and financed with state, municipal, and private funds. Using original documents Uta Rättig (1974) has shown that according to their statute, these primary schools were to instruct 8–10-year-old children from all social classes in a 5-year program of Mongolian, math, geography, history, and sports. The graduates could then attend a middle school with seven grades in the capital, where a faculty partially composed of foreign teachers[16] taught additional instruction in Manchurian, Russian, English, and French. For the first time in the history of education in Mongolia, study abroad was organized by the state. A dozen Mongolian students attended schools in the Siberian town of Irkutsk and in the Russian-Mongolian border town Khiagt from 1913 until the Russian October Revolution of 1917. Toward the end of the period of autonomy, efforts to create a national education system were intensified. In 1919 the newspaper *Niislel Khüreenii sonin bichig*, for example, called for sending all 6–7-year-old boys and girls to school, and for making the entire population literate in the next two decades (Rättig 1974: 489–495).[17]

In this third era of educational import, as in the first, the borrowing process entailed bringing in foreign teachers and sending students abroad. The new international orientation meant that foreign language acquisition became a central focus. Unlike in the previous two (and the following two) eras of import, there was no single foreign language that dominated higher education and served to indicate a leading reference society. The reason for this is both simple and paradoxical. To distance themselves from China and secure sovereignty multiple languages were emphasized. Thus foreign language acquisition ultimately functioned to preserve Mongolian culture.

Making access to education independent of social status, as was the practice in the first and second import eras, was carried over into the third as well. But this approach was now fixed in writing and explicitly represented as an attempt to supplying secular education to the entire population, thereby creating the conditions for a modern state. For the first time in Mongolian history, educating girls was officially considered desirable. This idea remained, however, at the level of "policy talk" (Cuban 1998) and according to Shagdar (2000) was never really put into practice.

The role of the press was among the most notable new features in this era of educational import, as Europe became evident as a reference horizon. For the first time in Mongolia, newspapers and magazines appeared, representing a completely new medium in the public sphere. In 1913 Tseveen, a Buriat strongly influenced by European culture, established the periodical *Shine Tol'*, which is still in publication today. The paper understood its main task as an educational mission: its content was intended, according to Baabar (1996 and 1999), to inform everyone who was literate about world events in a language they could understand. In 1915, Prime Minister Sain Noyon Khan Namnansüren initiated publication of the weekly paper *Niislel Khüreenii sonin bichig*, following a trip to Russia where he was impressed by an image of himself in a newspaper. Unlike *Shine Tol'*, however, *Niislel Khüreenii sonin bichig* printed articles explicitly on the topic of educational policy. Of particular interest was a description of a trip across the country by the prime minister, during which he arranged for children from the Barga area to start school. This can be understood as an instance in which education policy helped integrate regions of Inner Mongolia into the larger agenda of pan-Mongolism. A 1916 edition of the paper reported on modern European innovations in teaching methodology such as excursions and class trips making their way into the curricula (Jigmedsüren and Baljirgarmaa 1966).

In publications addressing the history of education an internal reference horizon can be identified, in addition to international orientation and the external reference horizons: repeated emphasis on Mongolia's own historical past. Shagdar (2000) mentions that numerous articles in *Shine Tol'* on foreign and domestic history as well as historical novels were published that, as he describes it, promote patriotism. We find that these reflections on the country's own history were related to two significant factors. First, they served the function of mental decolonization as the special accomplishments of the Mongolians themselves were highlighted in the context of a cultural renaissance. Second, this type of reflection served to reinforce their identity and sense of self in the nation building process.

In general, a euphoric sense of new possibilities dominated at the government level as well as educational institutions in the capital during the period of autonomy. But this mood did not carry over to the rural periphery, where there was a lack of innovation, shortage of faculty, and closure of many of the newly founded *khoshuu* schools due to low attendance. The monastic schools continued to exist during the Mongolian autonomy parallel to the secular schools. Their importance as a nationwide institutionalized network was even greater during this period, given that in the 747 monasteries there was a total of 120,000 lamas who either taught or studied in them (Jigmedsüren and Baljirgarmaa 1966: 3).

Following Schriewer (1990; see also Luhmann 1990), we can identify three legitimization strategies as the context for the third era of educational import. First, the self-referential recourse to traditions and values served primarily as a reaffirmation of identity during the process of decolonization, yet was also meant to provide continuity and stability. Second, the necessity of modernization for successful nation building was justified by an appeal to scientific rationality. Thus this legitimization strategy cannot be considered self-referential, as it was in the previous case. Third—and this is typical in the context of social transformation—additional meaning had to be created via externalization due to massive pressure for reform. Referring to "the other" meant reference to the "world," since education policy goals were aimed at internationalization, or "modernity." Most importantly, and this is what makes this case interesting, "internationalization" served the purpose of preserving Mongolian identity. It provided a way of referring to their tradition—vital, given the threat of China. The threat scenario also explains the choice of "Europe" (mediated by Russia) as the reference society. In the arguments employed by agents of educational policy, reference to history, stemming from the vision of pan-Mongolism, was accompanied by reference to the "world." Externalization was understood as a means to an end. As interesting as the borrowing patterns were in this third era of import, there is little more that can be said about the level of implementation because Outer Mongolia's autonomy, established in 1911, was already over by 1918–1919 with the invasion of Chinese troops.

Fourth Era: Universal Access

The Chinese occupation of Outer Mongolia was politically the most decisive factor in the country's choice of the Soviet Union as a new partner. The historical events that subsequently led to the Mongolian People's Revolution in 1921 cannot be discussed in detail here. However, when the repercussions of the postrevolutionary civil war in the Soviet Union spread across Siberia and into Mongolia in 1920–1921, Mongolia was caught in the "cross-wires of the military and 'world revolutionary' interests of the Bolsheviki" (Barkmann 1999: 185). The enemies of the "Red Russians" (Mongolian: *ulaan oros*)—the "White Russians" (Mongolian: *tsagaan oros*)—marched into Mongolia and drove out the hated Chinese. At first they were welcomed, but then a reign of violence began under the leadership of the Baltic baron Ungern-Sternberg. The Mongolians were only able to free themselves with the military and logistical help of the "Red Russians." Once this had happened, the country became interesting to the Soviet Bolsheviki ideologically, because they had effectively brought "the 'world revolution' into another country" (Barkmann 1999: 207).

Starting in 1921, Mongolia was increasingly oriented toward the Soviet Union and became the second socialist country in the world. Bulag expressed the ambivalence of Mongolian–Soviet relations when he described the pro-Soviet orientation as an "essentialized identity" motivated by opposition to China:

> Mongolian nationalism during the socialist period was characterized by a tension between a desire for development towards a Soviet-oriented civilization

and the wish to develop a national culture. The traditional identity was being transformed into the concept of a socialist "new Mongol." (Bulag 1998: 16)

This quote points to the definitive reference horizon of the fourth era of educational import. It also refers to the intended transformation of Mongolians into "new human beings"—a task to be accomplished primarily through education. It was here that the tension described by Bulag between dependence and attraction was manifest. This explains why shortly after the political changes in 1990, Mongolian education specialists began heated public debates about whether or not their countrymen had "lost their sense of being Mongolian" (*mongoloo aldlaa*) under Soviet influence, or if they had preserved their Mongolian traditions (*mongolyn yos oo*) (Erdene-Ochir 1991: 8). S. Bayasgalan has pointed out that even before the changes of 1990, critical voices questioned to what extent one can speak of a Mongolian school given the disavowal of certain national characteristics (Bayasgalan 1990). In an interview with the largest national newspaper in 1991, the Mongolian Minister of Education N. Urtnasan said that 70 years of socialist ideology had created an "illness," causing people to think only in terms of black or white (Urtnasan 1991: 2). Meanwhile, education expert and teacher N. Kausylgazy wrote an article in the party newspaper *Ünen* saying that in the education sector everything was presented as cut and dried, and that school structure, content, and changes in curricula had been adopted unquestioningly from the Soviet Union (Kausylgazy 1990).

These examples illustrate how harshly the recent past was judged in 1990. Mongolia was no different in this respect from other former socialist countries, where similarly radical judgments initially determined debate. In Mongolia, as elsewhere, the tone soon changed and angry verdicts gave way to more complex evaluations. One example can be found in an article from 1997 written by the current education minister, Ch. Lkhagvajav. Looking at educational import, he compared the positive and negative aspects of what he referred to as "Sovietized education" ("*zövlöltjsön bolovsrol*"). In his opinion, Marxist education philosophy was right in maintaining that any person can in principle learn anything because this was the basic principle underpinning universal education (including education for women). On the other hand however, the liberal legacy of early Soviet pedagogy fell prey to Stalinist repression and gave way to bureaucratic propaganda. Nevertheless, Mongolian educational science had profited greatly, particularly from the didactical work of Soviet scholars (Lkhagvajav 1997). This statement reflects the ambivalent attitude toward the fourth era of educational import still prevalent today, which we now consider.

In the initial years after the revolution, the change in political paradigms was expressed primarily in programmatic communiqués. In 1921, the new Mongolian government founded its own department for school-related issues in the Interior Ministry (Mongol Ulsyn Bolovsrol, Soyol, and Shinjlekh Ukhaany Yaam 2001). The primary task of this department, where the Buriat Tseveen worked, was to open schools. At first the practice was to open very few new schools, and instead use the schools that already existed from the period of autonomy. In 1924, following the death of the theocratic head of state *Bogd Gegeen*, Mongolia's first constitution was ratified at the Third Party Congress. The government was restructured, and the department for school-related issues was transformed into the Ministry for the People's Education (*Ardyg Gegeerüülekh Yaam*), based on the Soviet model. The Buriat intellectual Erdene-Batukhan was appointed to be the first People's education minister.

It is a phenomenon of the postrevolutionary period that so many high positions in politics and education were occupied by Buriats. Tseveen and Erdene-Batukhan had already made their appearance during the period of autonomy, but after the revolution a classic pattern developed according to which Buriats served as mediators in cultural, political, and language-related areas and introduced a revolutionary style. As Mongolians and inhabitants of a settlement area largely on Soviet territory, they were simultaneously both insiders and outsiders, and thus they were destined to take on the role of mediators.[18] Many of them were active on two fronts, as Buriats tended to be fervent nationalists and representatives of pan-Mongolism, which was soon to clash with the revolutionary ideas of the Communist International (Comintern).[19]

How did the restructuring of the education sector begin? The first constitution established the right to free education for all as well as the separation of church and state (1924 ony khevlel). The relevant passage in the constitution indicates that the state claimed to be the highest authority in the sector of education, despite the fact that the monastic schools still had a much greater presence. A secular "people's education" cannot be said to have existed initially, given that in the first years there was a lack of state schools and teaching personnel. By the end of the 1920s there were only 25 state-financed and 89 municipally financed elementary schools, and according to estimates they housed only 1.5–3 percent of all children (ca. 1,000), while approximately 13 percent (18,955) visited monastic schools (Schöne 1973: 18–19, 25–27).

Like the first educational import, this one began with the establishment of academic institutions. At Tseveen's initiative the Scientific Committee[20] was founded in 1921, out of which the Academy of Sciences was developed in 1961/62 (Chuluunbaatar 2002). Contemporary sources indicate that until the second half of the 1920s, the research climate was not yet dominated by an ideology antagonistic toward the traditional sciences. On the contrary, it was, in fact, surprisingly open and pluralistic. Orientation toward the outside world brought a great fascination with comparative studies, which were aimed at the integration of divergent perspectives. For example, historical studies were to be undertaken on the basis of a comparison of Mongolian and foreign language sources; in the field of medicine, traditional Tibetan healing methods were to be compared with European methods; and in order to establish an observatory, lamas trained in astronomy were to be engaged with the expectation that they learn about European astronomy (Unkrig 1929; Ischi-Dordji 1929). Ambitions to integrate scientific tradition and modernity led to an attempt to relate Buddhist teachings to those of Lenin. Morozova (2002) reports that two brochures were published in 1926 with the support of the Scientific Committee, in which the materialist elements in Buddhism were identified and Buddha himself was posited as one of the founding fathers of communism.

The perceived necessity of social modernization inspired the Minister of the People's Education Erdene-Batukhan to travel to the Soviet Union and Western Europe in 1925 to become informed about their education systems and acquire modern teaching materials. In 1926 his ministry decided to send around 40 Mongolian students to Germany and France, where they enrolled in various schools and vocational centers (Ischi-Dordji 1929; Wolff 1971; Schöne 1997). This was the first occasion for student exchange initiated by the Mongolian People's Republic. It was motivated by the pressure to reform and intended to ameliorate the lack of experts in the economic, administrative, and cultural sectors. Germany was attractive because of its developed network of technical schools and was thus chosen as a main location for training.

Meanwhile, the political climate in the Soviet Union was radicalized under the influence of the Communist International. In the wake of these changes during the late 1920s a characteristic pattern of Soviet–Mongolian relations developed in which almost every change of political course for the "older brother" soon recurred in Mongolia. The climate of openness came to an abrupt end with the Seventh Party Congress in 1928, when it retrospectively— in accordance with Soviet political terminology—was defamed as a "rightist deviance" from the party line. One of the first consequences of this for educational policy was that that in the wake of the subsequent radical change of course—later called the "leftist deviance" (1929–1932)—all students studying abroad were called back early at the command of the Far East Secretariat of the Comintern (Wolff 1971). The radicalization however was actually aimed at the Buddhist educational institutions. A ban on entering monastic schools in 1930 (Sandshaasüren and Shernossek 1981) was followed by a law in 1934 prohibiting the teaching of religious content in schools (Shagdar 2000; Gataullina 1981). Finally, the end of the 1930s saw a Stalinist style destruction of monasteries. This was accompanied by the assassination of many representatives of the religious intelligentsia, while those who survived were suppressed and prevented from contributing their knowledge and skills to society.

The destruction of the monasteries had a disastrous effect on education, which lasted for years. This was exacerbated by the fact that initially there were no existing alternatives for realizing the constitutional right to universal education. Although state spending on education was at T 10 million in 1940— 7 times higher than spending in 1925 (Gataullina 1981: 57)—by 1940–1941 only 10–11 percent of all children were in secular schools (Bawden 1968: 380; Uhlig 1989: 405). Unlike the monastic schools, state schools in the early years did not offer room and board for students from nomadic families due to a lack of funds. This is one of the reasons why obligatory schooling was only introduced in the mid-1950s when there were finally enough boarding schools following a broad range of reforms. We return to this issue in chapter 9. For now we consider the level at which this programmatic goal of "education for all" was undertaken.

The first political step in this direction was establishing the constitutional right to an education, as mentioned earlier. One result of this constitutes a new feature of this era of educational import, namely the regular inclusion of girls. Shagdar (2000) considers 1924 to be the beginning of coeducation in Mongolia. The central focus in educational policy into the 1960s was, however, making the entire population literate. This began in 1925 with the founding of a department for adult literacy in the People's Education Ministry, which—as in the USSR—was also assigned the task of ideological agitation. Correspondingly, most activities took the form of campaigns to "liquidate illiteracy" (Gataullina 1981: 51). In addition to the youth association, which primarily fulfilled agitprop tasks, roles were played by wandering teachers (Schöne 1982), roving theater groups, and movie theaters (Ischi-Dordji 1929), as well as former soldiers promoting adult literacy (Morozova 2002). Realizing the "people's education," in a literal sense, was the highest priority of education policy during socialism, and was considered a measuring stick of progress. Dorzhsuren writes:

> During the years of people's rule Mongolia became the first Asian country to eliminate illiteracy. For this great achievement the country won an high UNESCO award, the Nadejda K. Kroupskaya Gold Medal. (Dorzhsuren 1981: 109)

Historians of education usually refer to script reform as a catalyst in this context: In 1941 the Central Committee decided to replace the classical Mongolian script with the Cyrillic alphabet, with the addition of two auxiliary letters (Shagdar 2000). This followed a brief period of experimentation with the Latin alphabet, as has also taken place in Soviet Central Asia. This decision was justified with the argument that Cyrillic was more appropriate and would help speed literacy efforts (Gataullina 1981) and/or that the new alphabet would make it easier to equip Mongolian printing presses with Russian machines and make learning Russian easier (Schöne 1973). Even if the latter two arguments seem reasonable, they were not decisive. Instead, the primary goal was probably to use script reform to isolate Inner Mongolia and repress pan-Mongolism. In any case, aid deliveries from the Soviet Union came nearly to a halt during World War II. It was not until the year 1945–1946 that the Cyrillic alphabet was taught in all Mongolian middle schools and adopted as the official alphabet.

The founding of the Mongolian State University in 1942 was considered a milestone in the development of the socialist education sector, and it is a milestone also from the perspective of educational import. Previously, many students had to finish their studies in the Soviet Union. There were practical reasons for this, such as the lack of Mongolian textbooks for the natural sciences. After the university opened, initially with three departments (human medicine, zootechnics, and pedagogy), the situation began to change. The state university was conceived according to the Soviet model, and its direct subordination to the Council of Ministers points to its political dimension. Until after the first Mongolian graduates finished their degrees in 1945, the instructors all came from the Soviet Union. During the entire Soviet period up until 1990, Russian and Marxism-Leninism were compulsory subjects.[21] This not only served ideological purposes but was also intended to promote internationalism within the socialist camp. Also characteristic of the socialist university system was its vertical structure. Like the vocational schools, the university churned out graduates according to a plan determined by the labor market. Only as many students or apprentices were allowed to register as were needed to fill projected jobs for the year following the time it would take to complete a standard course of study.

As with the first era of educational import, the language of the primary reference society became the language of scholarship. Along with the language and the structures of educational institutions, ideological and didactical concepts and content were also imported from the Soviet Union. Having command of the Russian language was indispensable for many (though not all) university fields because a large portion of the materials was not available in Mongolian. Even after the Mongolian People's Republic had their own universities, postsecondary and vocational schools, Russian still played a central role. Not all educational opportunities were available in Mongolia, and there were always Mongolians studying or receiving vocational training in the Soviet Union. But it should be emphasized that Russian was *never* the obligatory language of instruction in elementary or secondary schools. Not only the Mongolians but also the Kazakhs,[22] who formed the largest national minority, were taught in their native language at least until they finished secondary school.

The structure of general schools (*Yerönkhii Bolovsrolyn Surguul'*) had been developed during the 1940s according to the Soviet model of 4 + 3 + 3 (4 years of elementary school, 7 years for incomplete, and 10 years for complete secondary school with qualifications for entering postsecondary education), and remained in place until the beginning of the 1960s (Sandshaasüren and Shernossek 1981: 26). In 1963 the Great People's Khural[23] decreed the "law on consolidating the connection between school and life and on the further development of the people's education system." This somewhat cumbersome title, as well as the content, were largely modeled on similar Soviet laws from 1958 (Shagdar 2000). The law called for expanding obligatory schooling to seven years, and later extended to eight. Meanwhile the secondary schools, like their Soviet predecessors, were to offer 11 grades with polytechnic and vocational training. But the schools were overwhelmed, and after two years of experimenting the plan was abandoned in favor of the old system (Gataullina 1981). The structure of the general schools was nonetheless expanded again in 1965 to 4 + 4 + 2 and starting in 1973 to 3 + 5 + 2. Despite repeated attempts the schools were never again expanded to 11 grades during the socialist period. This was also the case for the repeatedly stated goal of lowering the age of school enrollment from eight years to seven, or even six.

These waves of expanding and retracting goals are in some respects typical of school reform in socialist Mongolia—and the post-socialist era as well. Up until 1990, attempts to imitate the Soviet Union often came up against limitations because conditions in Mongolia could only partially be compared with the situation of the "older brother," and thus the implementation of desirable reforms often never took place, as we have just seen. These kinds of failures resulted partly from overzealousness and partly from ignorance of the local conditions. Of course, the importation of Soviet concepts did not always remain stalled at the level of policy talk or policy action (Cuban 1998), and there were in fact successes and partial successes in the implementation of reforms. Let us examine three examples.

In the first, and most successful, case (1) only an ideal was imported, but for its implementation a whole new path was taken. The two other cases are examples of partial success. The way in which they were presented utilized strategies of (2) simulation and (3) mimicry that were typical for socialism, but which are also still used strategically today in the fifth era of educational import.

1. The most significant accomplishment of the Mongolian education system during the fourth import era was without a doubt the realization of universal access to education. The socialist ideal and goal of establishing the "people's education" (Mongolian: *ardyn bolovsrol*) was implemented through various means. Two political conditions initially contributed to this, both of which were also characteristic for other socialist countries. First, all levels of education were free of charge, and second, the education system was highly accessible at all stages. Preschool was also free of charge. It was subsidized partly by the state and partly by the employer—whether the work was in a factory, a state ministry, or a cattle-breeding cooperative. This is because one of the political goals, beyond establishing preschool education, was to support working women, who benefited from the relief provided by preschools.

The real challenge for the Mongolian People's Republic, however, was achieving inclusion of the nomadic population. But for this, no readymade concept was imported beyond the general claim to offer "education for all" on the basis of Marxist educational philosophy. Realizing this goal in practice meant first and foremost adapting to local conditions, that is, integrating the factor of nomadism into education policy strategies. What did this mean concretely? Starting in the year of 1955–1956, obligatory schooling was introduced and the age of school enrollment was stipulated as eight years. In 1957 a total of approximately 94,000 children (92.8 percent of school-age children) received either formal or nonformal education across the country. In 1957–58 the school enrollment rate for children between the ages of 8 and 12 was 97.7 percent (Schöne 1973: 83, 103, 111; Shagdar 2000: 150–152, 182). Such a steep rise within such a brief period was not, of course, simply due to the introduction of compulsory education. It can also be traced to three external factors that simultaneously led to a fundamental restructuring of Mongolian society: first, the collectivization of livestock farming; second, the expansion of rural infrastructure; and third, a massive increase in external financial aid.

In brief, the influence of these three closely related factors can be described as follows: Collectivization in the 1950s changed the organization of labor in cattle-breeding families and released children from work. With the modernization of animal husbandry, risk was minimized and social security was increased, which meant not only that the threat of poverty from the loss of cattle was diminished, but also that individual lives could to a certain extent be planned. The effect of this on creating equal opportunity for people from rural areas in relation to those from the cities should not be underestimated. The boarding schools were the most important factor in the education system for the development of rural infrastructure. In the initial years the boarding schools were built using funds from the cattle-breeding collectives (Mongolian: *negdel*) and, starting at the end of the 1960s, primarily with funds from abroad. The rise in foreign aid was due to the Mongolian People's Republic becoming a member of the socialist trade association CMEA or Comecon, which we explore in greater detail in chapter 4.

The centralized organization of socialist society made it easier to coordinate the different subsystems and made mutual integration possible. It is characteristic of the Mongolian case that education policy was closely coordinated with the requirements of nomadic livestock breeding. The most important precondition was undoubtedly state financing of the boarding schools for all levels of education. The relatively late age of school enrollment (eight years) was not only a concession to the difficulty of providing room and board, but it was also a concession to the harsh climate. Vacation was organized according to the seasonal peaks of labor involved in pastoral animal husbandry. In addition, students were required to perform temporary agricultural work in the context of several different programs (sponsorship brigades, polytechnic instruction, summer and harvest camps, etc.).

We understand the motivation behind these mutually integrative policies not so much as a tribute to the nomadic *way of life*, but rather as an economic concession to nomadism as a *way of production*. But this is ultimately irrelevant in terms of the results attained. It is not without reason that the results were seen, especially from a comparative perspective, as a success for nomadic education unparalleled anywhere in the world. In addition to identifying crucial political decisions, Saverio Kratli's comparative study found a "non-antagonistic culture towards nomadism" at the level of implementation (Kratli 2000: 48). It was this "soft" factor that fundamentally distinguished the Mongolian model from the way in which nomadic education was realized in other states, including its reference society, the USSR. Attaining "universal access to education" was imported as an ideal, but the ways in which nomadic education was implemented were largely Mongolia's own creation, the success of which can be explained not least by the prudent integration of local conditions.

2. Let us now consider a counterexample, polytechnic education. In this case a deeply embedded socialist concept was imported. Gataullina (1981: 63) is correct in describing "polytechnicization" as one of the leading ideas of Marxist-Leninist pedagogy. Polytechnic education meant introducing instruction in various forms of vocational labor, starting at the secondary level as a way of applying the scientific or technical knowledge learned in school in a production-oriented manner (Njanday 1976). Since the 1920s in Mongolia, decrees calling for labor-related education had been issued at party conventions, but they had not yet been implemented. Shagdarsüren traces the difficulties in implementing vocational training back to the fact that Mongolia had bypassed capitalism en route to socialism[24] and thus struggled with the legacy of feudal backwardness, and in particular the absence of industry (Shagdarsüren 1976). According to Marxist-Leninist theory, as the name "polytechnic" implies, universal technical knowledge was to be imparted on a scientific basis. Further, the concept of polytechnic instruction was not limited to one specific subject, but instead the "polytechnic principle" was supposed to be integrated into all areas of instruction as a guiding theme.[25]

Courses in polytechnics were only made obligatory after passage of the new School Act of 1963. The subject was divided into three pedagogical areas: *moral* instruction aimed at conveying "respect for working human beings," *theoretical* instruction for imparting knowledge of polytechnics, and *practical* courses for teaching skills needed to operate machinery and instruments. In addition, advanced secondary students were taught the basics of technical drawing, mechanical engineering, and electrical engineering (Sandshaasüren and Shernossek 1981: 44).

Although the MPR, like other socialist countries, relied heavily on Marxist-Leninist guidelines for conceptual orientation and theoretical justification—citing in particular N. K. Krupskaya, A. V. Lunacharskii, and M. I. Kalinin (Jessipow 1971; Shagdarsüren 1976; Njanday 1976)—the promotion of polytechnics was an international phenomenon at the time. The theory and practice of polytechnics was also deemed desirable in the West (Lenhart 1993). Polytechnics can thus be considered a product of *zeitgeist*, gaining in popularity especially after the Sputnik shock. In the following years, the international reputation of socialist education systems grew. In addition to the praise for polytechnics, the promotion of gifted students in special schools targeting different subject areas, typical for socialist countries, also received attention. But this type of program did not become particularly developed in the MPR.

Starting in 1963 UNESCO invested $1,760,000 to build the Polytechnic Institute, which took over the task of qualifying technical experts (Schöne 1973: 160–162). Mongolia had an ambitious goal: the development of a modern industrial-agrarian state out of a pastoral economy. The enormous pressure to succeed and progress—which was felt at all levels of the system—was most apparent in the area of vocational education and training. One of the goals of polytechnic instruction was to steer students toward choosing an occupation, but success was only moderate. In a survey at the end of the 1960s, 36 percent of the 1,229 advanced students in rural and urban schools (ninth and tenth graders) who were asked what they wanted to become said they did not know yet. But of those who did have an idea of their future occupation, most of them wanted to work in nonproduction-related areas: as teachers (32 percent), doctors (15 percent), and drivers (11 percent). Shagdarsüren has identified two primary reasons why attempts to guide students failed. First, practical instruction was not based on up-to-date scientific and technological knowledge, and second, teachers often did not know how to integrate the "polytechnic principle" into their subject area (Shagdarsüren 1976: 93–96).

But what did the implementation of polytechnics look like in practice? A few successes could be found here and there. For example, school brigades in polytechnic courses built school benches and other objects for science instruction classrooms, or even yurt structures for external commissions (see Njanday 1976). School production brigades were also established in some agricultural cooperatives and their affiliated machinery stations (Mongol Ulsyn Bolovsrol, Soyol, Shinjlekh Ukhaany Yaam 2001). But nationwide implementation of more sophisticated concepts was significantly limited by the lack of industry in numerous locations. For many rural areas it can be said, in general, that polytechnic instruction had an agrarian character. For example, in 1979 the Council of Ministers decided that students in lower-level grades would be regularly involved in raising young cattle and in the higher grades would participate in the harvest and procurement of animal feed (Shagdar 2000). At the level of implementation, it seems that the purpose and goal of this wholly industrial-oriented concept was largely lost.

It was not only the practical realization of this project that was Mongolized but also the theoretical bridging of the gap between theory and practice. One strategy was to lower expectations. Njanday (1976), for example, declared polytechnic instruction in livestock farming a success by taking into account general aspects such as the knowledge gained, the potential for building character, and the transformation of nature. In addition, there were also attempts to find parallels between real existing practices and Mongolian traditions. Shagdarsüren (1976), for example, discovered numerous correlations to the way in which traditional work practices were passed on in cattle-breeding households. He thus justified the displacement of the concept of polytechnics into the completely different context of cattle breeding in a way that is both seductive and elegant. For him, socialist polytechnics seemed a logical and worthy heir to traditional popular education. To reinforce his argument, he makes reference to Krupskaya, who noted that as industry is being transformed, agriculture also gradually takes on a socialist (i.e., industrial) shape (Shagdarsüren 1976: 12).

The Mongolization of polytechnics is a vivid example of the simulations typical of socialism: Since almost no similarities with the original idea of polytechnics could be found, they simply behaved as if they existed. This was possible in practice by performing rituals oriented toward something that did not necessarily have to be present, given that everyone involved had an idea of what it was and were all well versed in simulation techniques. In the domain of theory, it became a ritual to transform inadequacies into successes by means of the proper interpretation, which meant devising clever redefinitions until connections to the original concept could be established.

3. Let us look at a last example for the Mongolization of imports that is also exemplary of the wave-like repetition of reforms: The idea of "making instruction more effective," continued today under the label of "student-centered learning,"[26] was attempted again and again. The import of didactical concepts from the Soviet Union had been apparent from the very beginning. Works by Amonishvili, Shalatov, Lysonkova, and Karakovskii, for example, were present over the course of a teacher's training. But this pedagogical import was intensified due to a crisis. In the context of the new School Act of 1963, experts began to evaluate the quality of instruction at general schools. The goal was to make content and structures better conform to modern needs and develop means for securing progress in learning. But studies done by the education ministry from 1961 to 1966 yielded unpleasant information. Particularly in the basic subjects of Mongolian and math, student achievement was far from ideal; the number of students failing was high, particularly for grades 4 and 5, while 23.7–36 percent of students from schools with 4 grades left school early. Also rather troubling was that students were not capable of working independently. In a study in 1968, the pedagogical department of the Academy of Sciences diagnosed serious methodological inadequacies in many of the teaching staffs. Studies showed that teachers used only 0.8 percent of classroom instruction time to promote independent learning (Tsevelmaa 1965: 31–36; Schöne 1973: 130–132, 152; see also Eggert 1970: 414–415; Gataullina 1981: 65).

As a result, overcoming student "passivity" was considered the greatest pedagogical challenge over the next two decades and became (and still remains) a point of emphasis in reform efforts (see Begz 2002). In 1966 a commission was charged with evaluating curriculum content according to didactical criteria and formulating new programs for specific subject areas. Parallel to the commission's work, a system for the continuing education of teachers was developed, and development courses subsequently became an integral part of the teaching profession (Schöne 1973). In addition, the spectrum of pedagogical journals targeted toward practitioners at various levels was extended.[27] School No. 1 in Ulaanbaatar was chosen as a laboratory for the Pedagogical Institute's experimental trials for new initiatives. The teaching staff at the school developed new methods for making students more active and encouraging them to work with greater independence during the learning process (Gataullina 1981). Starting in 1972, all curricula were based on the methodological principle of "developing teaching," the main goal of which was to motivate students to think and work on their own (Sanshaasüren and Shernossek 1981; Shagdar 2000). In higher education there were attempts to compensate for inadequacies by establishing preparatory classes for outgoing exchange students.

Students' lack of ability to work independently remained a persistent problem and many explanations were suggested. Uta Schöne pointed to organizational shortcomings and a traditional student–teacher relationship in which students were not urged to take initiative. Their "studies" were instead often limited to repetition of material read aloud by the teacher (Schöne 1973: 137). This is apparently an example of a persisting "residuum" (Steiner-Khamsi 2003) that has its origin in education traditions typical for the culture. In the earlier discussion of the first era of educational import, we addressed elements of the traditional student–teacher relationship and the importance of rote learning. It is certainly plausible that these traditions again and again triumph over imported concepts because they occupy a high position in the culturally determined hierarchy of values. The teacher is thus expected to be the subject who knows, while the student is expected to reverently strive to acquire their knowledge.

The evaluations led to concerted efforts to solve the problem, which were partially successful. But not all improvements claimed for the system were real. A typical strategy was what could be called mimicry: In order to demonstrate better figures, Kausylgazy (1990) found, bad grades were simply given less often. There were even cases in which degrees were handed out for courses of study where only some of the requirements were taught. It was also popular to "solve" problems by changing the categories used to assess them. These kind of strategies flourished during socialism, and as we see later in case studies (chapters 7, 8, and 9), they are reenlisted today when necessary. Mimicry patterns developed because the issue at stake was often not as important as the timing of the report itself: scrutiny by authorities and anniversary events created pressures to prove success. Not all imports were welcomed at the level of implementation, which is also why no "ownership" feeling was developed. Today, however, it is just as important to please the decision makers.

In conclusion, we have seen that for the fourth era of educational import significantly more new features can be identified than in the three preceding eras. The establishment of a modern school system with universal access to education in a nomadic pastoral society is doubtless unique. In this sense the Mongolian People's Republic forged new paths as they implemented the imported concept of the "people's education"—paths for which no role models could be found, not even in their reference society, the USSR. Most of the other new features (preschools, a multitiered system, polytechnics, vocational training, the establishment of the modern sciences, a vertical structure in higher education, internationalized graduation requirements, and degrees) were, however, based almost exclusively on the Soviet model.

In 1990 the socialist era ended for Mongolia. That same year the International Education Conference ratified the "Education for All" program for the coming decade. It is one of the ironies of history that the education system in Mongolia simultaneously began a decade of decline in which the prized feature of "universal access" would be seriously undermined.

3

BYPASSING CAPITALISM

In an image known to everyone in Mongolia, often found adorning walls or book covers or stamps, a fearless Mongolian rider carries the national flag atop his white horse. Both horse and rider seem to leap out toward the viewer (see photograph 3.1). The powerful visual dynamic is due to the spatial perspective and the forceful forward movement. Foreground and background are linked through the horse's jump, its long mane flying in the wind alongside the huge flag. They come galloping out of the dark and after a broad jump over a barrier, they proceed into a bright red foreground. But what is it they have jumped over? Every Mongolian knows the answer: capitalism, of course! A closer look reveals a banner written in Cyrillic capital letters which reads in Mongolian as "*KAPITALIZMYG ALGASCH*"—"BYPASSING CAPITALISM." This image was created in 1959 by the Mongolian artist Dagdangiin Amgalan (born in 1933) during his student days at the Moscow School of Art to commemorate the thirty-fifth anniversary of the Mongolian People's Republic. It represents at a single glance the entire Marxist-Leninist theory of history in its adapted Mongolian version.

For anyone familiar with Marxism-Leninism, the symbolism of the image is immediately apparent. The horse and rider have left the dark feudal past behind them, boldly jumped over the (equally dark) stage of capitalism, and landed square in the bright center of socialism. The banner marking the "bypassing of capitalism" is sufficient, the rest is self-explanatory. The waving flag boasts the Mongolian national colors—red and blue—which the rider holds as if whisking them away from the dark past in order to bring them safely into the light. Although dressed in traditional clothes—signaled by the Mongolian hat with red earflaps flying in the wind and traditional boots with pointed upward curling toes—the rider fulfills all the iconographic criteria of a true revolutionary hero. His gaze looks forward (into the future), he is holding up the national flag (and the destiny of his country), and he has the reins in his hand (i.e., he knows the way).

His horse is pure white, which is not only meant to be striking but also to symbolize speed and drive, as well as valor and gallantry. White horses have always signified purity and exaltation, even the Chinggis Khan is said to have preferred white horses. As the "alter ego of the Mongolians" (Veit 1985), the horse of a hero is considered a magical animal that—according to many Mongolian epics—can easily climb to higher spheres and thereby move

Photograph 3.1 Bypassing Capitalism
Source: D. Amgalan (painting from 1959)

into other worlds. Thus, as Amgalan's image shows us, it is not an impossible feat for the horse to jump over a dark, exploitative social order and to land in a better world. As the artist later said in interviews—without horses, the Mongolian People's Revolution would have been unimaginable. The horse was the first thing that came to mind in far away Moscow when he was trying to find a suitable symbol for the Mongolian revolution (Tuyaatsetseg 2001).[1] The iconographic composition was suitable to the demands of socialist realism—nationalist in form and socialist in content. Horse and rider served as the ideal prototypes of Mongolian heroism, accomplishing a patriotic deed in the spirit of the Marxist-Leninist theory of history.

The Mongolian Version of Marxist Historical Theory

Karl Marx's historical inquiries led him to the conclusion that the development of human societies occurs according to predictable patterns. His basic premise, put succinctly, was that particular material preconditions bring about certain modes of production, which in turn shape the social structure of society. All this depends upon the given property relations of a society because property relations produce inequalities, or, as Marx described it, classes that exploit and classes that are exploited. Marx's other main historical thesis is that human productive forces continually work to improve the instruments of production, thereby acting as a driving force in progress. Material culture constitutes the base structure upon which intellectual development is appended, to a certain extent, as a dependent "superstructure." This is expressed most succinctly in the Marxist theorem: "Being determines consciousness" (Mongolian: *akhui ukhamsryg todorkhoilno*).

So how did Marx then explain the transformation from one social order to the next? In each historical phase he identified certain social groups that had been particularly successful and innovative in further developing and exploiting the means of production (i.e., production instruments and production forces). Because their movement forward was hindered by the dominant social relations, they worked toward transforming these relations into a social order that better suited their aspirations. On the basis of this idea, Marx formulated a model of social development that initially implied a four-tiered progression: primitive society, slave society, feudalism, and capitalism. But what was to come next?

According to Marxist theory, social classes developed because the means of production were the private property of a small group of people, enabling them to dominate over those who needed to perform manual labor to secure their livelihood. Another important Marxist theorem is thus, "Whoever owns the means of production maintains power." What disturbed Marx was his realization that the exploited classes were responsible for most of the social value produced, yet they were only able to claim a very small portion of it for themselves. He thus came to the conclusion that it was time for philosophers to stop interpreting the world and instead to begin to change it. Because he understood private property as the origin of class society and thus the cause of exploitation, he began to imagine ways in which private property could be eliminated, thereby creating a classless society free of exploitation.

Assuming there was little hope that the ruling classes would willingly give up their property, Marx appointed the class most exploited by the capitalist system to be the future agent of history, and assigned them the task of overthrowing the old order through a socialist revolution to transform private property into the property of the people. Marx intended socialism as merely a temporary transitional phase (!) during which exploitation would be stamped out and the way prepared for the real goal—the classless society of communism. It is well known that this "highest stage of history," although never reached, served as an imagined promise, providing the official view of history throughout the entire socialist camp with its meaning, goal, and orientation. After the end of socialism, the loss of this unabashedly prescriptive, yet plausible and seductive teleology was perceived by many as a forfeit—not only of a set of ideals, but also of the only form of historical orientation they knew.

This feeling of historical displacement was expressed in many of the interviews conducted by Swetlana Alexijewitsch, a journalist from Minsk, in the early 1990s. Alexijewitsch interviewed former Soviet citizens who had tried to take their own lives, as well as the family members of those who had committed suicide. Her book constitutes a substantial contribution to reconsiderations of the history of socialism, in large part because she does not objectify her subjects as "socialist people," or denigrate them as mere "products of socialism." Instead, she allows us to hear the voices of those who have reached a depth of experience unknowable to others, having been exposed both to real existing socialism as well as a proximity to death. For the author, there is no question that those who experienced socialism have the right to evaluate their present lives in terms of their past; indeed, this is a central tenet of postsocialist studies. The images of history prevalent in postsocialist nations today—including not only disillusionment with the past but their immanent projections of the future—must be compared with competing worldviews. One of Alexijewitsch's interview partners, a former speaker for the first secretary of the District Party Committee of the CPSU, expressed his thoughts as follows:

> Socialism forced the people to live in history, to make history. . . . It brought them together via an activity, a direction, uniting them. A grand idea kept the chaos at bay. . . . The people felt that they were carrying out a historical task, that they had a part in something great. . . . And what can you offer people? A life of satisfaction? Prosperity? For us, that will never be the final goal. We need a tragic ideal. (Alexijewitsch 1999: 99)

The idea not only of being involved in a historical process, but by necessity "living in" the midst of history and "making history" according to a guiding principle, was omnipresent under socialism. This idea was based on the Marxist model of successive social development mapped out earlier, and it gradually became what Uradyn Bulag (2002) has called, with reference to Foucault, a historiographic narrative. As such it had the function of translating politically relevant theoretical elements into clear worldviews that are ideologically sound, while simultaneously giving them a positive spin as logically consistent. In this context, Amgalan's representation of "bypassing capitalism" can be described as *the* central historiographic narrative of the Mongolian People's Republic. As such, it not only fulfilled various ideological functions during the socialist period, it is still present in political discourse in Mongolia today as a point of reference and debate.

Next, the milieu in which this narrative developed must be considered. Here it becomes apparent that the thoroughly materialist Marxist theorem "Whoever owns the means of production maintains power," has an idealistic complement. In this instance the power of definitions plays a key role: Definitions not only create new concepts, they also have the power to create new social facts. After 1990, in many debates in former socialist countries the statement could be heard that socialism (i.e., as an idea) could not live up to reality. In the case of Mongolia, as we demonstrate with our example "bypassing capitalism," the opposite could also (ironically) be said—reality could not live up to socialism. This is mostly because the construction of the historiographic narrative was an ambitious endeavor. Moreover, it is striking how the permanent capitulation of reality was transformed into an image of permanent victory through the use of ever new interpretive concepts. Let us start by considering the beginning, that is, the revolution itself.

The events of 1921—when the "Red Russians" came to the aid of Mongolians to emancipate them from the Chinese occupiers and the "White Russians"—were later defined in terms of the Soviet "October Revolution," as the "Mongolian People's Revolution." By definition, every revolution must necessarily replace the old social order with a new one. And this is where the historiographic problems began in the case of Mongolia. Let us first recall the Marxist ideal model of historical social progression from primitive society to slave society, through feudalism, to capitalism, socialism, and finally communism. Marx presented this model as universally valid and based on self-evident principles. For the revolutionary agents, the immediate goal of socialism was as clear as the ultimate goal of communism. What was not as evident was the starting point for Mongolia, given that at the beginning of the twentieth century capitalism had not yet been introduced. And the absence of capitalism was not the only problem. Also missing was an identifiable working class, the most important ingredient in the theoretical formula for socialism. Mongolian society, with an estimated population of less than 1 million at that time, was instead made up of princes, Lamaist monks, and various groups of nomadic cattle breeders. These facts presented the Soviet revolutionaries with the dilemma of how to resolve theoretically an undeniable contradiction. Ultimately, they could not afford to exclude potential partners in the struggle for worldwide socialist victory just because their social order was not (yet) "ripe" for the revolution.

Vladimir I. Lenin's question—What is to be done? (Russian: *Chto dyelat'?*)—was formulated as the title of a text he composed in exile in 1902, which later became required reading for revolutionaries (Lenin 1945). After the "Great Socialist October Revolution" of 1917, Lenin was considered the first thinker who had been able not only to realize Marx's revolutionary theory, but also to develop it further in a pragmatic direction. What were the implications of these developments for Mongolia? After the victory of the "People's Revolution" in Mongolia in 1921, Marx's theoretical model of history was posed with another challenge. Although Russia at the time of the October Revolution in 1917 could boast the necessary ingredients of capitalism and a proletariat required for the theoretical formula of revolution—at least on its European side, this was not true for the later Soviet republics of Central Asia.[2] As the leader of the Bolsheviks, Lenin could not let himself be held back by these facts, considering that the goal of the Communist International (Comintern)[3] was worldwide revolution.

> Lenin developed and updated it at a new historical stage in history and, in his report to the Second Congress of the Comintern in 1920, pointed out, among other things, that "with the aid of the proletariat of the advanced countries, backward countries can go over to the Soviet system and, through certain stages of development, to communism, without having to pass through the capitalist stage." (Shirendyb 1981: 20)

Lenin's creative exegesis of Marx's theory was in effect a high level authorization for the Central Asian Soviet republics, and Mongolia, to simply "bypass capitalism." This idea then became the essential historiographic narrative for Mongolia (Günsen 1962; Sükhbaataryn neremjit khevleliin kombinat 1964; Shirendev 1967; Sh.U.A.T.K. 1971; Shirendyb 1971, 1978; Vietze 1978; Gundsambuu 2002a). But what function did it serve, in terms of both representation and legitimization?

It is not incidental that the aforementioned quote refers to "certain stages of development." After bypassing capitalism, it still proved difficult to invent a theoretical foundation for defining the present and future social orders. The ideologues were faced with a sticky paradox: They were supposed to demonstrate the objective truth—and thus the inevitability—of Marx's teleological model of history (which Marx had merely ingenuously "uncovered") by using an example that contradicted all of its prescriptive prognoses. The idea that thanks to Lenin's creativity Mongolia would be spared an entire level of development—and an exploitative one at that—seemed like a good one, but how would socialism be erected without a working class and in the complete absence of industry? For orthodox Marxists and Comintern activists, only one answer seemed possible: Here the real existing social structure would have to be adapted to fit the theoretical model. It was clear that building up industry in Mongolia would take time, and thus it would also take time for a working class to be created. Who, then, should lead the class struggle?

According to Marxism, this is the task of the downtrodden and exploited masses. But in Mongolia, the search for appropriate agents in the class struggle was hindered by many factors. Not only were "the masses" missing, but the social structure had been shaped by nomadism and thus did not fit into any of the available theoretical categories. But, as had been the case with the model of progressive stages, this ultimately proved to be a question of definition. In the context of "Marxism or unilinear evolutionism" (Khazanov 1994: 13), it was assumed that the nomadic and sedentary societies must in principle develop similarly. Labeling the socioeconomic relations in Mongolia "nomadic feudalism" (Mongolian: *nüüdliin feodalizm*) was a logical step in demonstrating this. The term "nomadic feudalism" was originally shaped by the Russian Mongolist Vladimirtsov (1934), who was not really a dogmatic Marxist himself and whose neologisms first became part of orthodox Marxism only after his death. In any case, calling the old social order "feudal" still did not solve the overall problem because the next task was to identify the social classes in accordance with the theory, and then to assign them a particular role in the class struggle.

Let us briefly consider which social groups lived in Mongolia in the 1920s. There were aristocrats with various aristocratic titles (*noyod*), nonaristocratic cattle breeders (*arad*), Lamaist monks of various ranks (*lam*), and the monastery subjects (*shav'*). Within the *arad* group, some had their own cattle, while others, the *khamjlaga*, herded the cattle of the aristocrats and/or wealthy *arad* families. All *arad* were obliged to pay duties and perform civil service (*alba*). In contrast, the *shav'* who served the monasteries were freed from these duties, which meant that former *arad* often changed their status to *shav'* (Natsagdorj 1967; Barkmann 2000). At this point the ideologues ran into trouble again, first because according to the theory, class structures were supposed to run along the lines of property, and second, it was supposed to be clear who allowed whom to work for them, that is, who would be exploited. Once again the situation in Mongolia proved complicated. Among the *arad*, for example, there were considerable differences in the property they owned, and the wealthier let the poorer *arad* work for them, just as the aristocracy did. Meanwhile, the wealthy who wished to gain virtues gave part of their herds to the monasteries, who lent them to the propertyless *shav'* in exchange for the use of milk and wool (Humphrey 1978). Lending animals to the *khamjlaga* in exchange for their care was also a common practice (Natsagdorj 1967). In short, a clear separation between the exploitative feudal class and the exploited herders was impossible. As a consequence, it was difficult to define class struggle in Mongolian society because, as Bulag (1998) rightly pointed out, the modes of coexistence gradually eroded the differences between groups.

Apologists for the Marxist historical model, however, were unhindered. They came up with a new categorization for the Mongolian social structure according to the prescribed Marxist-Leninist criteria for class structures: the *noyod* and *lam* were designated the feudal exploiting class, and the *arad* (including the *khamjlaga*) and *shav'* were considered the exploited class. This solution allowed for class struggle by suggesting two antagonistic classes, thereby also designating the appropriate agents for constructing a socialist society. The nomadic cattle breeders in Mongolia, who could not be considered part of the feudal class, functioned as a substitute proletariat and were lumped together under the name "*arats*" (Mongolian: *ard*, from classical Mongolian: *arad*). This term later became a regular feature in literature on

Mongolia from the socialist camp. The creation of the umbrella term *ard* not only had an important function in labeling the historical agents, but also in founding the Mongolian People's Republic in 1924, when yet another terminological problem arose. As David Sneath (2003) has convincingly demonstrated, no semantic or functional equivalent existed in the Mongolian language for the Russian word "people" (*narod*). When it came to inventing an equivalent for such an important term, the word *ard* seemed closest in meaning: "the common people." But problems in constructing definitions did not end there.

Once the term "feudalism" was taken out of its original European context, the next stumbling block was establishing who held property rights to the land. Property was hard to identify in Mongolia because the main question with regard to land was always who had the right to use it (Natsagdorj 1967; Bawden 1968; Barkmann 2000; Sneath 2002a). But for the defenders of Marxist-Leninist theory, the concept of "ownership of the means of production" was such a central principle that they found an answer, and along with it a grounds for class struggle. Official historiography described a "class of nomadic feudal lords" who arose in the thirteenth century, that is, during the period of Chinggis Khan, and became the owners of the land and pastures. The cattle breeders, the "class of direct producers," were dependent upon them because they themselves owned no land (Zlatkin and Gol'man 1982). Khazanov (1994), one of the harshest critics of enlisting the concept of feudalism in relation to Central Asian nomadic societies, argued that this was inappropriate because it suggested similarities with Europe that were ultimately superficial. The fact that elements can be identified which could be characterized as "feudal" (such as dependence, e.g.) does not justify a Eurocentric classification under this rubric, according to Khazanov.

Regardless of Khazanov's trenchant critique, the idea of "nomadic feudalism" was advanced in socialist literature on the Mongolian People's Republic and became orthodoxy. For the discussion at hand, the function of this idea proves relevant to the "bypassing capitalism" narrative. A review of the relevant publications (e.g., Sodnomgombo 1978; Vietze 1978; Sandshaasüren and Shernossek 1981; Schinkarjow 1981; Zlatkin and Gol'man 1982; Picht 1984; Jügder 1987; Barthel 1990) yields a strikingly unified picture. The descriptions of feudal relations—which were, as we have seen, in large part a presumptuous construction—fulfill three ideological functions in these texts. First, they serve to contain nationalism; second, they glorify socialism; and third, they serve as a model case with international significance.

First, let us consider an example from the area of education. In 1949 the Central Committee of the Mongolian Revolutionary People's Party decided upon a far-reaching curriculum reform in accordance with the dictates of their sole paradigm of interpretation—a revision of all history and literature textbooks to represent the Marxist perspective. This reinterpretation was so extensive that even Chinggis Khan was marginalized in the texts, having been "revealed" as a member of the exploitative feudal class.[4] The (historically brief) postrevolutionary period comprised at least half of the lessons in history (Schöne 1973). This example illustrates that the revision was intended not only as an anchor for the Marxist viewpoint, but also as a way of devaluing national history. The socialist concept of patriotism subsumed national history under the banner of proletarian internationalism, that is, the national question was considered a subordinate part of worldwide class struggle (Lchamsüren 1978). Any expression of nationalist pride that made recourse to the prerevolutionary period, or elevated individuals not qualified as members of the "the people" was deemed nationalist, that is, reactionary and separatist. It was only during the period of thaw after the death of Stalin that Mongolian intellectuals could successively rehabilitate their national cultural heritage (Damdinsüren 1959).

Second, the descriptions of feudal relations in the relevant literature had the function of delineating the economic and cultural backwardness of prerevolutionary Mongolia and then contrasting this with the accomplishments of socialism. These passages almost always began with statements such as the following:

> Ever since its foundation the Mongolian People's Revolutionary Party has been carrying out the cultural transformation of society, originally a society of almost total illiteracy and utter religious fanaticism which for centuries had vegetated under feudalism and serfdom. (Dorzhsuren 1981: 109)

As this example shows, "illiteracy" was the central metaphor for the backwardness of feudal relations and was deemed the "heritage of feudalism" in Marxist historiography (Sanzhasuren 1981: 89). Socialist education thus became the measuring stick of progress and combating illiteracy became a favored motif in various, contrasting representations. But this motif not only served the task of self-representation to the outside world, it functioned internally as well in terms of before-and-after comparisons used in periodic literacy campaigns to promote reading and writing among the adult population. For example, the administrators of the Choibalsan and Khentii provinces addressed a letter to analphabets in 1961, saying:

> In the first year of the cultural campaign there were 1,389 illiterates in our *aimag*. Now the number is 250. But you are among their numbers and you are thereby increasingly left behind in our rapid development.... Before the Revolution only a few people could read and write, now we have several doctors and scientists.... Imagine, there are space crafts that are sent to the moon 405,000 kilometers away where they circle around and take photographs. There are radios in the herders' yurts and elsewhere in the far steppes that can receive radio programs over distances of a thousand kilometers. There are also cows with a yearly milk yield of 12–13 thousand liters, sheep that produce 15–20 kg of wool, and draught horses that can pull 50 t, and these are the result of the constant effort and diligent work of intelligent people. You have not yet been able to read about these accomplishments.... if you cannot read, it is the same as being blind. (Schöne 1973: 306–307)

Third, The Mongolian People's Republic was considered the second socialist country in the world and the first on the Asian continent. Although considered to have an atypical development from the Marxist point of view, together with its big brother the USSR, the Mongolian People's Republic had taken part in founding the socialist world-system. We have seen how the atypical was incorporated into, and became an affirmation of, the official line thanks to the "bypassing capitalism" narrative. Later an opportunity arose for representing the case of Mongolia as a living example that verified the theory. In the early 1960s when newly independent colonies, particularly in Africa and Latin America, had gained their independence, both sides of the Iron Curtain sought to sway them in their direction. These Cold War battles provided Mongolia with the opportunity to make its grand entrance as a flagship of socialism. Mongolia was presented to the so-called backward countries of the "Third World" as a representative example of a grandiose development, an "ascent out of medieval backwardness and feudal oppression to socialism" (Shirendyb 1981: 19). In consequence, the "bypassing capitalism" narrative took on a new dimension in public representation to the outside world. This in turn evoked a new self-understanding within the Mongolian People's Republic of being an essential part of the socialist world-system due to the country's unique history. In chapter 4 we further discuss the implications of this.

Climbing Up and Down the Stairs: Teleology and Never-Ending Transition

How could the transition to a new order be explained in theoretical terms, particularly given that socialism could not yet be said to exist in Mongolia? Lenin's notion concerning the bypassing of capitalism referred to "certain stages of development." This was not coincidental. In order to be called "socialist," a country had to meet certain requirements. Although the Mongolian People's Republic had counted as the second socialist state since the 1921 revolution, it was only at the Tenth Party Congress of the Mongolian Revolutionary People's Party in 1940 that the "creation of the foundations for the development of socialism" (*sotsializmyn ündsiig baiguulakh üye*) were declared part of the official agenda, thereby inaugurating the transition to state socialism. Why did this declaration come so late? Before a country can call itself "socialist," not only did a minimum level of industrialization and a certain level of development have to be established, but the "exploiting classes" could no longer be in existence. And it was precisely this last factor that was considered accomplished in Mongolia at the end of the 1930s following waves of Stalinist purges. Using the example of Inner Mongolia, Bulag (2002) has convincingly shown that the potential for violence was inherent to the classification of "people" and "landlords." Certain groups in the Mongolian population became primary targets of the class struggle, and there were no available means for controlling the use of this discursive concept. In the Mongolian People's Republic at the end of the 1920s and the beginning of the 1930s, thousands of lamas and aristocratic and nonaristocratic owners of large herds fell prey to a radicalized interpretation of the

Marxist model of history, which condemned them—under Stalinist influence—as the "exploiting class."

In the official Mongolian historiography, of course, the brutality of this process was not addressed. Rather, as was typical with socialism, a division into the different "stages" (Mongolian: *shat*) was described, with implicit reference to Lenin's statement concerning "stages of development." In general, this was undertaken retrospectively and had the apparent function of locating a place within the Marxist prescriptive teleology for even atypical examples. The historical epochs with their various labels were not only part of the curricula for history lessons at all levels of instruction, they were also present as a deterministic paradigm in almost all other subjects. Through prayer-wheel-like process of repetition, this form of historical positioning became immanently present to everyone. These representations of history make it clear where the feeling quoted earlier in Alexijewitsch's (1999) interviews came from, namely the sense that socialism induced people to live inside history and to feel united in a predetermined historical course. But what were the labels used to define the historical course of the Mongolian People's Republic?

According to Charles Bawden (1968), Mongolian historians divided the postrevolutionary period until 1940 into the following three epochs: (1) the phase of consolidation and preparation for the declaration of a People's Republic (1921–1924) (2) the phase of debates leading to the general party line (1924–1932), and (3) the subsequent course along a noncapitalist line of development (1932–1940). The literature unanimously points to the year 1932 as a turning point, generally in reference to the "liquidation of the economic power of the feudal class." In contrast, Bawden himself (rightly) understands the year 1928 and the Seventh Party Congress as a decisive turning point on the basis of the fact that this was when Mongolian politics were largely synchronized with the Soviets. In any case, in later publications with a Marxist-Leninist bent, the entire period between 1921 and 1940 was labeled "antifeudal" and/or "democratic stage of the revolution" (Shirendyb 1978, 1981; Gataullina 1981; Zlatkin and Gol'man 1982; Bormann 1982; Barthel 1990). This yields two results that are valuable in helping us understand the Mongolian variant of communist historical theory and also the patterns of interpretation still common to this day: first, the connotation of the word "democratic," and second, the recognizable tendency in descriptions of these historical phases to characterize each as a transitional phase of the underlying teleology.

The appearance of the words "(generally) democratic" in this context may come as a surprise to many Western readers. This label was intended to demarcate a mere (and lowly!) preliminary stage leading into a subsequent socialist stage of development. In Marxist-Leninist theory it was considered necessary and legitimate to secure the revolutionary change of power following the initial overthrow of the ruling class through violent means (Norovsambuu 1971). For this purpose, a temporary "dictatorship of the proletariat" was necessary to bring to power those social groups who had previously been identified as part of the "oppressed class." The underlying understanding of democracy at play here is based on the idea of erecting a "rule by the people." The "people" referred to were the propertyless (Sanjdorj 1971), and thus the removal of the "antagonistic exploiting class" was interpreted as a deeply democratic act that justified brutal means (Tsedenbal 1954; see also Dashzeveg 1971). In the Mongolian People's Republic where, as we have seen, the lack of a proletariat in the postrevolutionary period had to be painstakingly integrated into the theory, an adapted official version of this can be found in the program of the Mongolian Revolutionary People's Party: ". . . a democratic dictatorship by Mongolian labouring people had gradually evolved into working-class dictatorship" (cited in Gundsambuu 2002b: 5). Independent of attempts to find legitimization for the country's origins, the citizens of the Mongolian People's Republic harbored an omnipresent sense that they were living in a democratic state (Günsen 1962), and in fact the structures of school administration did show some democratic traits (Steiner-Khamsi and Stolpe 2004).[5]

As for the second point, the apologetic use of the Marxist model of historical stages forced a teleological inflexibility onto interpretation of these phases that left no room for anomalous realities. The labeling of special sub-epochs was a welcome aid here: The labels were intended to legitimate the party's political line, which was to appear as an inevitable necessity, with no alternative given to the immediate goal of "socialism" and the long-term goal of "communism." In many cases the *ex-post* labeling made it clear which obstacles had already been overcome on the route to a brighter future. This device was an effective means for justifying deficits and "Parousian delays" (Riegel 1993: 336). In effect, the invention of the various "stages" did not remain limited to classifying society as a whole; soon it flowed over into a variety of historiographic representations—from the collectivization of cattle breeding (Rinchin 1981) to combating illiteracy (Gataullina 1981). Because history, understood teleologically, could not be seen as an open-ended process, it was necessary to lend the various epochs a transitory character by designating them as such. This also provided an explanation for the retarding moments experienced under the status quo. The problem was never attributed to the prescriptive theory; it was instead the "lack of socialist consciousness" among the people that slowed down progress.

The earlier mentioned decision made at the 1940 Party Congress, declaring that the creation of the "foundations of socialism" should become part of the agenda effective immediately, meant that a transitional phase was once again underway. The task was thus clear: the indispensable ingredients for socialism, "industry," and "the working class," were to be "created" (Tüdev 1971; Nansal 1971; Shagdarsüren 1976). In a country whose economics were dominated by nomadic pastoralism, the education system was made responsible for accomplishing this,[6] and it was only in the 1960s that socialism was considered to have been more or less "established" in Mongolia. During the period from 1961 to 1980, the republic entered the stage of "creating the material and technical base of socialism" (Shirendyb 1981; Sükhbaataryn neremjit khevleliin kombinat 1964) with the goal of transforming Mongolia from an "agragrian-industrial into an industrial-agrarian country" (Pelzhee 1981). From then on, whatever the new epochs were called, Mongolia found itself alongside all the other socialist countries in a never-ending state of transition (Mongolian: *shiljiltiin üye*). This manifested itself everywhere in the infinite absence of the promised communist paradise. In all socialist states during this period of permanent transition, absence was justified by the invention of ever new substages. Michael S. Voslensky has described this pan-socialist approach as follows:

> everyone cheered the notion that communism would be established in the Soviet Union by 1980; now everyone cheers because an unexpected "developed socialism" has evolved out of what was established, though it is still a long way away from communism. (Voslensky 1984: 477)

The experiences gathered by individuals during socialism, interpreted historiographically, also structure the perception and evaluation of events since the political changes in 1990. In Mongolia many jokes were made in reference to the "bypassing capitalism" narrative: Now we have to catch up on the capitalist phase we skipped back then, which just goes to show that you will be punished if you don't follow the Marxist model. Or: It's just like at school, if you miss a lesson, then you have to repeat the class. Or: We just took a step backwards, but which way is forward from here?

The irony of history is that this question already had an answer waiting. In place of the old formulaic phrases so often repeated in socialism, there was now a new mantra for the future agenda: "TransitionToFreedomDemocracyAndMarketEconomy." Where this new path was headed could be seen in the many examples available in the "West." History was once again not an open-ended affair, the goal was crystal clear and needed only to be reaffirmed. And in a familiar way the same premature propaganda songs of praise for the glorious coming future could be heard across the former socialist countries. Caroline Humphrey (2002b: 21) aptly noted that "The current 'transition' being attempted in Russia is heavily ideologized or mythicized, no less than the revolutionary transition to socialism." Labeling former socialist states "transformation countries" reveals that the new interpretation of history was based on a narrow understanding that, as Henning Schluss and Elisabeth Sattler (2001) point out, was not much different from the mechanistic vocabulary of electrotechnics. In electrotechnics, transformation implies that the final state is known before it has been reached and with the help of certain physical laws can be determined and helped along on its way via technical means. Coincidental interactions or even the input of active agents are not imagined to play a role here.

The concept of transition and transformation countries introduced after the end of socialism represents a new "teleology of transition" (Verdery 1996: 227) that once again mapped out a schematic course of development for the states of the former Eastern Bloc. The pattern of deterministic positioning in the course of history was all too familiar, and something else also created a sense of déjà vu: After a few years of progress along the new course of "transition," more or less according to the prescriptions imposed by international financial organizations, the promised ideal—this time defined as a prospering market-oriented democracy—has once again failed to materialize. In Mongolia today, in the year XVI,[7] with growing poverty and enormous social differences, there is no doubt that this transitional state will remain the status quo. Given their previous experiences with political prophesies, disillusionment arrived even before the new prescriptive transition concept was discarded as an inappropriate and externally imposed label. Khayankhyarvaagiin Gundsambuu (2002b: 142) is in agreement with most Mongolian academics when he asserts that ". . . Mongolian economic life is atypical for economic laws both of socialist and capitalist structures, if not totally free of their influence." But the uncritical transfer of concepts has also met with harsh criticism in the West among experts on Mongolia. Sneath has studied certain aspects of the effects of the recent changes on nomadic cattle breeders. Introducing an institutionalist perspective in his study of the new market ideology, Sneath asserts,

> In some respects the conceptual shift from a "feudal" to a collective notion of property can be seen to have been a less radical change than the one proposed at present, as the government attempts to introduce a market economy. (Sneath 2002a: 200)

Continuing his analyses, Sneath observes the meaning of privatization in the Mongolian context:

> Collective Mongolia adapted old notions of a hierarchical unitarian society, which was well adapted to the organization of pastoral life. In the current "age of the market" we see that privatization neither means the same thing as it does in the West, nor does it produce the same results. Old networks continue to be used, and there are numerous "price distortions." The behaviour of pastoralists does not conform to the laws of supply and demand. (Sneath 2002a: 204)

The ignorance of real existing peculiarities is thus as common to both images of history as is the deterministic disavowal of history as an open-ended process. Both views also share a common answer to the question of who is responsible for the delay in reaching the ideal state: Earlier the population lacked the necessary "socialist consciousness," now they are accused of harboring a "socialist mentality" and/or, as Humphrey (2002b: xx) has found, they are represented absurdly, as obstacles to introducing a market economy. This inadvertent comedy of eternal repetition suggests that, in the terms just introduced, reality will be as unable to live up to capitalism as it was to live up to socialism.

CONTEMPORARY REPRESENTATIONS OF THE COMMUNIST STAGE MODEL

Whether thwarted by reality or not, the Marxist model of the historical stages of development has not been forgotten; indeed, it is manifest in the present context as a symbolic pattern of interpretation. And the good old concept of feudalism is also still a favored discursive point of reference. One example of this can be found in Gundsambuu's sociological study of herders as the largest group in the Mongolian population, wherein the author takes a critical view of the politics of transition pursued since 1990 (Gundsambuu 2002b). According to him, the dissolution of the collective led to the destruction of the rural infrastructure, thus bringing the living conditions of the herders back down to a level they had at the beginning of the twentieth century. The advantages of democracy and a market economy are hardly tangible in rural regions, he claims, which is why "some people even perceive this transition as a drawback and regression leading to nostalgia about the old socialist society" (Gundsambuu 2002b: 134). What is meant by "drawback" is a presocialist, that is, "feudal" period and thus a time when Mongolian herders did not have access to the structures that constitute modernity.

The rhetorical recourse to feudalism as *the* symbol of backwardness is one of the most popular rhetorical strategies used today to critique the present. This is less a symptom of a purely pro-socialist position than of a skeptical stance toward prescriptive prognoses for the future and the accompanying disavowal of catastrophic mistakes. Just as party politics were legitimated under socialism via the labeling of various stages, the "transition" terminology represents neoliberal structural adaptation as a necessary inevitability. While Marx's laws served earlier as "objective" and thereby unquestioned absolutes, the "law of the market" has been introduced as its ideological successor. In the current discourse of postsocialist countries, the use of the term "feudalism" is a metaphor for, as Humphrey has poignantly expressed, a "vision of hopeless entrapment" (Humphrey 2002b: xxii) in a life of poverty that leaves no room for the idea of freedom and which cannot be reaffirmed as democratic progress by those affected. On the basis of her research in postsocialist Romania, Verdery (1996: 228) has even established that for these reasons the present is in fact perceived as a transition from socialism to feudalism and creates "images of a backward shift in time." In chapter 9 we explore the catalysts in Mongolia's education system that spurred these perceptions of a slide back into premodern relations after 1990 by looking at the example of nomadic education.

Finally, at this point we would like to suggest that Amgalan's pictorial representation of the "bypassing capitalism" narrative fits the description of what Arjun Appadurai (1997) has defined as an ideoscape, that is, an image component of the way the world is imagined. Ideoscapes can be invested with different meanings depending on the context. While "bypassing capitalism" was integral to the state ideology until 1990, it was not then replaced, but rather recycled to function in opposition as a negative critique of the present system. Amgalan's horse and rider have remained relevant, however what they represent has changed.

In the 1997 *Human development report Mongolia* (Mongol Ulsyn Zasgiin Gazar and UNDP 1997), a drawing by a child appears on page 51 in the Mongolian version (page 43 in the English version—Government of Mongolia and UNDP 1997).[8] In the drawing, a young Mongolian man wearing the traditional national costume amidst the yellow-green steppe is seen jumping forward over a dark barrier that crosses the image diagonally. In the background, little red flags display letters spelling out the word "*SOTSIALIZM*" (socialism), inscribed across the steppe. In the foreground, red and green dots form the words "*ARDCHILSAN NIIGEM*" (democratic society). The dark barrier the man is hurdling over is riddled with black cracks and the word "*YADUURAL*" (poverty). On second glance it becomes apparent that this barrier is in fact a deep trench. On the edge of the hole are two small human figures waving their arms in the air, as if calling for help. This time the hero has no horse, he is alone in overcoming the obstacle. His jump is not high and looks rather hectic—his tongue hangs out of his mouth, his hat is falling from his head, and one foot looks as if it could get caught in the hole to bring him tumbling down. Unlike the revolutionary hero, this young man has no air of heroic optimism, his gait makes him look as if he were on the run rather than on a mission. The cracks in the earth around the hole representing poverty suggest that it is expanding—and getting harder to bypass. The barrier has now become an abyss, into which one may fall en route to the future. This drawing represents a creative play on the composition and symbolism of Amgalan's work. It is hard to say whether the child consciously intended to represent the postsocialist transition from one social order to the other as a jump downhill (backwards?). In any case, the "bypassing capitalism" narrative is clearly still present here, having undergone its own transition since 1990—with its outcome open-ended.

4

EXCHANGING ALLIES: FROM INTERNATIONALIST TO INTERNATIONAL COOPERATION

As the second socialist state in the world—and the first in Asia—the Mongolian People's Republic (MPR) worked with the Soviet Union to form the foundations of the socialist world-system. The political constellation of these two countries—both "encircled by hostile capitalist states" (Luvsanchultem 1981: 136)—was considered the beginning of international relations "of a new type, based on the principles of proletarian internationalism" (Kazakevich 1978: 196). Since the end of World War II, the MPR had been a part of the socialist world-system and it became a member of the Eastern Bloc Council for Mutual Economic Assistance (CMEA) in the year 1962.[1] Despite its relatively late economic integration into the socialist community, the MPR never doubted its standing as part of the Second World. This identity was a product of the "bypassing capitalism" narrative, as well as pride in having a relatively well-developed social system, compared to other countries in Asia. It was also bolstered by the role of the MPR in socialist discourse on development, and the country's educational system was considered an indicator of its progress. In context with the decolonization of Africa and Latin America in the 1960s, the MPR served as a role model for the Third World, and promoted itself as an example of the opportunities offered by socialism for successful development from "feudalism" to modernity, and a higher standard of living.

The termination of the CMEA in 1991 was a catastrophe for the Mongolian economy. Industrial production and the gross domestic product diminished by a third, an energy crisis forced many companies to stop production, and real income was reduced by half within a year (Odgaard 1996: 116). Being newly classified as a "developing country" was a shock and an insult to the Mongolian people. They felt stigmatized for being ranked with "classic" developing countries in Africa and Latin America. This chapter explores why this drop in status from Second to Third World country was so painful for Mongolia. We also examine the change in alliance that followed, from an internationalist to an international paradigm of cooperation, in terms of its implications for educational development.

Younger Brother—Older Brother

With the creation of the socialist world-system following the World War II, the era of socialist internationalism began. This period was considered a qualitatively more advanced stage of proletarian internationalism, the guiding principle of cooperation between communist parties and nations. Realization of the "historical mission of the working class," that is, the spread of socialism/communism across the world (Picht 1984) was the goal. The point of reference was the Soviet Union, and further points of cohesion were the Marxist-Leninist worldview shared by other socialist countries, as well as resulting political and socioeconomic commonalities. Economic cooperation solidified through the CMEA was intended to bring the countries closer together through a gradual alignment of growth rates (Tichomirov 1978). In order to effect multilateral as well as bilateral trade relations, the CMEA created its own financial institution, the International Bank for Economic Cooperation, which conducted its finances by calculating the currencies of its member nations into transfer rubles. Cooperation among member nations was subject to a division of labor that aimed not only at specialization, but also regional diversification through the coordination of national economic plans in what was called the Complex Program.[2]

After a preparatory phase, the MPR joined the CMEA as a full-fledged member in 1962. Within 7 years 95.1 percent of its foreign trade was to be with the other CMEA member nations (Harke and Dischereit 1976: 10). In the period from 1958 to 1970, Mongolia received 1.5 billion rubles in economic aid from the CMEA, and from 1966 to 1970 40 percent of all investments were financed through foreign loans (Tichomirov 1978: 413–414). Other socialist countries with higher levels of productivity and faster growth in national incomes understood it as a "responsibility" and "social priority" (Rathmann and Vietze 1978: 346) to provide "fraternal" support for the Mongolian PR, and gave preferential treatment in the form of loans and special export prices (see Mostertz 1982). "Brotherhood" was the central metaphor of socialist internationalism, and this meant equality within the "family." In the Mongolian language, however, there is not only a lexical but also a semantic difference that distinguishes younger and older brothers, who are granted special rights and duties according to the principle of seniority. The creation of the socialist world-system meant that in addition to an older brother (*akh*), that is, the Soviet Union, Mongolia also had many younger siblings (*düü*).

But the Mongolian People's Republic did not have the economic capacity to act as an older brother to its siblings, and this was the most important reason for the country's late integration into the CMEA. Instead, it was taken under the wing of other socialist brother countries (*akh düü sotsialist ornuud*) and given substantial aid, without ever being labeled a developing country. There were three main reasons for this, which we consider consecutively. First, it would have reflected poorly upon socialism if a country that had been socialist as long as the MPR was not yet on the same level of economic development. Second, political priorities in the socialist camp were focused primarily on human capacity building, independent of the real level of economic growth. And third, the MPR's exemplary role as a flagship (and older brother) in the context of decolonization, was more important than its economic standing in the CMEA.

First, the problem of Mongolia's relatively meager economic capacity was rarely addressed publicly by the CMEA. In the instances when it was mentioned, the "bypassing capitalism" narrative served to explain Mongolia's special role as an "agrarian-industrial state" that had yet to be transformed into an "industrial-agrarian state." But it was also desirable to present Mongolia—in accordance with CMEA statutes—as an equal partner, and this was done using rhetorical tactics of evasion. The relevant publications contain charts comparing the gross domestic product of selected socialist and capitalist nations. Here one finds the absolute figures for the founding members of CMEA and the German Democratic Republic, but missing are any references to the Mongolian People's Republic. These are found instead only in those charts listing the relative rates of growth, because this is where Mongolia's economy looked good on paper. According to the statistics, Mongolia had a growth rate of 10.4 percent in its industrial gross domestic product in the period from 1951 to 1967, which corresponded to the average for CMEA countries at the time. For the developing countries the average growth rate was 8.3 percent, while the capitalist industrialized countries showed a rate of just 5.2 percent (Markow 1968: 751–755), making them seem worse off than they actually were since the starting figures went unmentioned.

Also absent from the official historiography of the CMEA are the economic advantages reaped by the MPR in the power struggle between the USSR and China. This factor, along with the selective presentation of statistics, distorted economic perceptions of the MPR. Uradyn Bulag writes that the initial "honeymoon" (Bulag 1998: 15) between the USSR and China ended at the start of the 1950s once Mao asked Stalin when he intended to honor the agreement made in 1923–1924, that Mongolia would be given back to China if Mao was victorious in the Chinese Civil War. Stalin did not answer, but the Chinese persisted. Subsequently, both the Soviet Union and China entered into a "bizarre competition" (Bulag 1998: 15) in which they lavished the MPR with aid. This rivalry was clearly to Mongolia's benefit, however it was also evident that aid from China was underpinned by threat and contingency. In a referendum in 1945, the Mongolian people had voted for independence from China in foreign affairs, and they now sought the protection of their "older brother." This led to further animosity between the two powers as well as the eventual exodus of most Chinese skilled laborers in the 1960s. Mongolia's entry into the CMEA offset the negative effects this would have otherwise had.

Second, during socialism, education and social policies took priority regardless of the real rate of growth because these areas were considered the motor of development. The following quote shows how the internationalist development paradigm was diametrically opposite to the later premises of "transition" in 1990:

> It would be unrealistic to try and solve the manifold and complicated problems of diversification solely on the basis of market mechanisms. . . . Many tasks—the training of work forces, the development of new sectors and new branches—cannot be accomplished on the basis of cost effectiveness. (Andreassjan and Eljanow 1968: 749–750)

This paradigm of socialist development oriented primarily toward human capacity building clearly had advantages for the MPR. The enormous successes in the education sector would have been unthinkable without it, as would the gradual modernization of the national economy. Mongolia's success in making the entire population literate, achieved by the 1960s despite numerous difficulties, commanded international recognition. The MPR received the Krupskaya gold medal at the 1970 UNESCO convention in Tehran for being the first Asian country to "eliminate" illiteracy (Dorzhsuren 1981: 109; Sandshaasüren and Shernossek 1981: 12).

After Mongolia joined the CMEA, the educational system received massive material and personnel support. Education centers, particularly for the purpose of professional training, were created according to Soviet models. After a decree by the CMEA Council of Ministers in 1964, numerous skilled laborers came from other socialists countries, often serving as vocational instructors (see Amarkhüü 1968; Kunzmann 1981; Dashtseden 1984). They were not, however, considered development aid workers, rather they were called "specialists" who offered "brotherly help." The language of communication was most often Russian, the lingua franca of the internationalist world. Many young Mongolians took advantage of the educational exchange efforts and completed their training or university degrees in other socialist countries. During the period 1961/62 to 1970/71, 280–462 Mongolians received training abroad in accordance with treaties for educational cooperation, and 1,553–2,666 studied abroad (Ardyn Bolovsrolyn Yaam 1976: 20–23); in the 1980s the latter figure grew to over 6,000 (Dashtseden 1984: 9–10). Within the socialist camp, to a certain extent, attempts were made to create internationally standardized graduation requirements in order to insure that students' degrees were compatible for study abroad.

In elementary and secondary education emphasis within internationalist cooperation was on money to train teachers and build new schools to keep up with rapid growth in the student population, particularly in the 1970s. According to Shagdar (2000: 204), in 1975 almost a quarter of the entire population was attending school. In the period from 1970 to 1977, 180 modern school buildings with classrooms equipped for science instruction and 161 additional boarding schools were erected (Sandshaadsüren and Shernossek 1981: 49–50). From 1970 to 1976, many former primary schools were transformed into schools with eight grades, and the number of schools hosting ten grades also increased across the country (Shagdar 2000: 206). In the larger towns, most schools had high numbers of students. For example, the No. 3 Middle School in Darkhan, the second largest town in Mongolia, had 1,300 pupils, and schools in the capital city counted between 1,600 and 2,000 pupils (Eggert 1970: 417–418).

Beyond their cooperation in terms of financing and personnel, there was also extensive academic collaboration between the CMEA countries. Close academic ties existed between the Pedagogical Research Institute of the MPR and its counterparts in "brother" countries, as evidenced by numerous conferences, research projects, and faculty exchanges. The influence of Marxism-Leninism as a guiding ideology and a structuring perspective within academia also led to commonalities in pedagogy with other "brother" nations, the declared goal of which was to bring forth the "new human beings of socialist society" (Dondog 1974: 132). Within the classroom this meant educating students in proletarian internationalism. An example of this comes from one of the most frequently translated standard texts for pedagogy:

> By the fourth grade the pioneers are usually already corresponding with children from other socialist countries. They write letters to them describing how they study, what they do in their free time and what kinds of socially useful work they do. They write about their parents' work and send their friends books, photo albums and various souvenirs. (Jessipow 1971: 202–203)

In addition to pen pals, meetings were held between students of socialist countries in international pioneer camps.[3] They admired the same heroes from the pantheon of socialist idols in politics and literature, and shared everyday experiences such as holidays and work weekends (Russian: *subbotniks*).[4] All of this led to a cultural homogenization within the socialist world. Beyond the official values that were taught, a culture of shared language and jokes evolved which allowed for an understanding across borders, and through which vague references could be decoded and meanings read between the lines. Proletarian internationalism in the classroom also included creating bonds of solidarity with nonsocialist countries in the struggle to free the world from capitalism. Whether in the GDR, Czechoslovakia, Bulgaria, or the MPR, all children who attended socialist schools in the 1970s were engaged in the "fight" to free communists imprisoned in other parts of the world—such as Angela Davis in the United States and Louis Corvalan in Chile—by writing postcards addressed to foreign heads of state.

Third, socialist development policy was understood as "internationalist solidarity." As the colonial system was crumbling in the 1960s, national liberation movements in developing countries received substantial aid; they were considered a third arm of world revolution after the workers' movements in capitalist countries and the socialist world-system. The powerful systems of the "First" and "Second World" were engaged in a bitter tug of war, trying to pull the countries of the "Third World" into their own camp. The development policies of the two camps were contrasted using the polarizing rhetoric of the Cold War:

> Some Sovietologists try to compare the Soviet Union's aid for developing countries with the so-called "support" policies of the U.S., which, as we know, are meant to serve the military and political interests of the U.S., . . . and to strengthen the financial, economic and political dependence of the "Third World" countries on the US and to keep these countries within the sphere of influence of the capitalist world system. (Natsagdorj 1978: 42)

It is this competition that brought the MPR its moment of glory as the flagship of successful development under socialist auspices. Simultaneously, due to the "bypassing capitalism" narrative, the opportunity finally arose for the MPR to appear as an experienced older brother in relation to other nations where the appropriate social order was not get in place:

> The non-capitalist path of development is a specific form of social progress. This path, which lies before the backward peoples of the world, is now history for Mongolia, a bygone stage in their development. (Sanjdorj 1978: 471)

The MPR also finally achieved membership in the United Nations in 1961, which they could use as a platform to speak in favor of other developing countries. In the tug of war between the two systems, the MPR representatives appeared as self-confident guarantors of a successful socialist development policy (see Shirendyb 1978) and proudly emphasized that Mongolia, which had until recently been a "backward country," was now in a position "to help countries which have taken the road of independent development" (Agvaan and Bat 1981: 152). The country's international influence grew extensively, in accordance with its role as a potential policy lender. At the end of the 1970s the MPR was said to have established official relations with over 40 former colonies in Africa, Asia, and Latin America (see Rathmann and Vietze 1978: 352).

Cuba and the Democratic Republic of Vietnam were the first countries to come under consideration as potential recipients of Mongolia's fraternal care. Communalities with these countries were immediately apparent: Cuba had similarly been the first socialist country on its continent, and North Vietnam, like the MPR, had also stepped over the "capitalist stage of development" (Nguyen 1978: 13). In its role as older brother, Mongolia soon had a relationship with these other countries not just at the diplomatic level, but also in classrooms and among pioneer organizations:

> The young Sühe Baatorists[5] [sic] are great internationalists. . . . In solidarity with the children of heroic Vietnam, Mongolian children carried out a nationwide campaign for peace and friendship and sent the Young Pioneers of Vietnam red ties and other gifts. (Galsan 1981: 127)

What remained the decisive factor in Mongolia's role as a flagship of socialist development was and still is their success in the area of education.

And their pride was justified. In 1989–1990, at the end of the socialist era, the literacy rate was at 96.5 percent (Mongol Ulsyn Zasgiin Gazar and UNDP 1997: 7), giving Mongolia the right to proclaim "a higher literacy rate than the United Kingdom or the United States, and a higher tertiary education rate than most countries in the developed world" (Mongol Messenger 1997: 7). In this context, it becomes apparent why the country was hit so hard by its demotion to the status of developing country in the early 1990s. According to the criteria of "gross domestic product" and "economic diversification," Mongolia was considered a least developed country (LDC). The relatively high standard of living, especially in terms of life expectancy, literacy rate, and school enrollment ratios, was irrelevant to this new categorization. Furthermore, the new label was so contrary to the Mongolian people's perception of who they were, that in 1993 the government rejected the LCD ranking, feeling that it was a discredit to the nation (Odgaard 1996).

Although the country felt categorization as a "Third World" country was a stigma, the Mongolian government was forced to accept this new label due to their dependence upon foreign loans. The "internationalist" aid at the end of the 1980s was 30 percent of the GDP, and the "international" development aid, in the form of cosponsoring from individual nations and multilateral lenders, was 25 percent in the mid-1990s, that is, about the same (Bruun and Odgaard 1996: 26). What was decisive, however, was not the volume of aid, but the shift in political and ideological orientation.

Donor Logic in International Cooperation

Although the tie between the "older brother" and "younger brother" suggested a lifelong union in common cause for the coming communist world revolution, the bond began to loosen in the mid-1980s,[6] and was completely dissolved by 1990. After protest and public unrest, the Mongolian People's Revolutionary Party (MPRP) conceded to a multiparty system and agreed to form a transitional coalition government. The transitional government (1990–1992) prepared the revision of the constitution, and paved the way for the first free elections in 1992, wherein the MPRP was rewarded for its reform-minded course. The demise of the CMEA and the Soviet Union not only cut the country off from vital economic assistance and trade, it also left Mongolia with huge debts toward the Russian Federation. In 1989, the amount of debt was established at $10 billion, which translates into the sizeable amount of $5,000 per citizen. Although the big debt was permanently settled in January 2004,[7] it overshadowed a relationship that had purportedly been one of solidarity and kinship.

Since 1992 the two political camps have regularly taken turns in forming the administration all four years; twice in landslide victories of the oppositional parties (1996: Democratic Union, 2000: MPRP), and once with a very thin and contested majority (2004: Motherland-Democracy Coalition). The landslide victory of the Democratic Union (1996–2000), followed by a similarly large ratio of voters changing their mind in favor of the MPRP (2000–2004) has triggered numerous speculation. The practices of stakeholder replacement, favoritism, and social re-stratification along party lines, with politicians generously rewarding their ardent supporters, are hinted at and joked about. It is said that each citizen who is left behind empty-handed at the end of each election, gives their vote to the opposition party in the next election hoping for some compensation.

Mongolia's dependence on external assistance has remained extremely high, with the country ranking fourth on a list of 30 countries that are classified as "aid dependent" or "highly aid dependent" (Ulziisaikhan 2004). Not surprisingly, each administration seeks to top the previous administration in racking up a greater amount of foreign aid, especially in the form of grants. There is a strong belief among politicians, government officials, and citizens that loans should only be signed if they generate employment and increase tax revenues. The period during which the country was receiving no donations was brief. Nevertheless it contributed to the economic collapse of the early 1990s. Already in 1991, the transitional government established the first agreements with the IMF, World Bank, ADB, Japan, and 14 other countries (Batbayar 2003: 51). Japan pledged $55 million in the early transformation period, and the country has to date remained the largest donor in Mongolia. Batbayar (2003: 52) asserts that one-third of the total aid for Mongolia flows from Japan in the form of grants, loans, and technical cooperation programs. The second largest source of external assistance is the Asian Development Bank, followed by the World Bank whose contribution represent about 11 percent of all loans (World Bank 2004a). Germany has been the fourth largest player in the field of international donors. The United States, although currently ranked fifth, vowed to assume a greater role as a donor and solicited proposals to be provided by the U.S. Millennium Development Fund. From 1991 until 2004, the Government of Mongolia received $2.6 billion in external assistance (half in loans), heavily concentrated in the economic sector, for transport, industry, construction, electricity, and heating (World Bank 2004a).

International donors in Mongolia pursue specific priorities and strategies that have little to do with the specific situation in the country itself, but rather mirror their own mission and objectives, systematically pursued in all of their so-called target or priority countries. It is therefore essential to understand donor logic and reflect on how it impacts policy borrowing and lending. We distinguish between four types of donors operating in Mongolia: international financial institutions, bilateral donors, the United Nations system (UN), and international nongovernmental organizations (NGOs).

International Financial Institutions

It must be emphasized that the IMF, World Bank, and Asian Development Bank are first and foremost banks. Development, for the World Bank and ADB, is of only secondary importance. Thus they behave like banks and their greatest concern is that they be repaid for their loans. That said, interest on the loans is not the primary concern in Mongolia,[8] but rather the conditions that must be complied with in order to receive the loans. For example, the most recent ADB loan has a maturity of 32 years, including a grace period of 8 years (ADB 2002: iii). The annual interest is only 1 percent, and therefore it's more a grant than a loan. The important difference, however, is that as a loan it must eventually (in 40 years) be repaid. Moreover, the conditions of the loan are strictly enforced and these conditions have an impact on education policy.

Receiving a loan or a grant from an international financial institution entails that education reforms be linked with economic growth. When economic concerns are allowed to play such a dominant role in development, many vital education reforms fall through the cracks, and numerous other initiatives are, as is be discussed in chapter 5, ill-conceived. Reducing public expenditures, generating additional income from households and businesses, and developing an educational system that is cost-effective are the top priorities for international financial institutions. Another problematic feature is tight international control over the Mongolian government. A myriad of reasons are put forward to rationalize why the debtor is not to be entrusted with developing, implementing, and evaluating a reform project on its own. The reasons range from accusations of "unprofessionalism"—an effective justification for putting governments and local experts in the backseat (see Escobar 1995)—to waste, nepotism, and corruption. As a result of this strategy of disempowerment, government officials and local experts are reduced to pawns who execute orders given by international consultants. The corollary is that when no orders are given, little gets done. Demanding subservience is therefore not only irrational and inefficient, but also expensive. Thirty percent of all loans and grants that the Mongolian government has received have been spent on "technical assistance" (World Bank 2001), that is, on international experts sent to Mongolia for short-term missions. The lack of any serious attempt to reduce dependency on these consultants is ironical given the rhetoric of national ownership, cost-effectiveness, program efficiency, and local capacity building among international donors (Eade 2003). Unfortunately from a comparative perspective (McGinn 1996) disbursing one-third of loans and grants on international consultants is not unusual in the "development business." Even worse, incentives for subservience and passivity contribute to the popular stereotype of Mongolians as quintessentially lazy. The myth of laziness is a trope that is perpetuated in Mongolian literature (e.g., Gundsambuu 2000a, 2000b), permeating everything from the interaction of class monitors and regular students in schools (see chapter 6), to encounters between international and local experts. Similar to other aid-dependent countries, there is enormous public pressure on the government to sign fewer loans and attract more grants, assume the role of a "key player," "stand up on its feet," and curtail the cost of "foreign consultants" (Ulziisaikhan 2004: 2).[9]

Both of the two international financial institutions operating in Mongolia pursue their own Country Assistance Strategy (World Bank) or Country

Assistance Plan (ADB), in which past experiences are evaluated, and priorities for a next phase of external assistance are outlined in detail (ADB 2000; World Bank 2004a). From these strategic plans we learn whether any financial assistance to the education sector has been scheduled for the upcoming years. Among international donors the ADB is particularly vocal and visible in Mongolian education reform for one main reason: it closely advises the Ministry of Education on what to do, and what not to do. It has been instrumental in developing all education sector reviews and education sector strategies on behalf of the Ministry of Education. This cooperation may be characterized as too close for comfort (see Edwards and Hulme 1996), in that the lines have become blurred between imposed and voluntary reform. It is clearly a conflict of interest if one and the same institution (ADB) prepares a needs assessment, writes up a (sector) review, develops a 5-year sector development strategy, appraises the project cost, and then makes the government sign a 40-year loan for a sector development project that the bank itself has initiated. As a result, sector reviews in Mongolia are neither very analytical nor terribly comprehensive, rather they are loan-directed. Formulated with a tone of crisis and calls for immediate action, the sector reviews simply carve out niches within the education sector the ADB has decided to loan money to. The target audiences for education sector reviews—all written in English—are the Board of Directors of ADB and the Ministry of Finance of Mongolia. This audience must be convinced that a loan earmarked for education rather than for finance, transport, energy, or other sectors is worth the investment.

As of 2005, two comprehensive multiyear education sector development projects and four small technical assistance grants have been agreed upon by the government and ADB, notably the Education Sector Development Project (ESDP) from 1997 to 2000, and the Second Education Development Project (SEDP) from 2004 to 2007.[10] The total project cost for ESDP was $15.15 million, and the cost for SEDP is estimated at $68.5 million (ADB 2002: appendix 5).[11] Although ADB appears to be focused on education in Mongolia, its allocation to the social sector (health, education, social insurance reform) is a meager 9 per cent of its overall financial commitment to Mongolia (ADB 2000: 7). Chapter 5 deals explicitly with the structural adjustment component ("rationalization") of the ESDP, and chapter 9 briefly comments on the expectations from the SEDP in the area of rural school development.

The monopoly of the ADB on comprehensive education reform in Mongolia has put the Ministry of Education in a strange bind. It acts at the mercy of a bank (ADB) when it decides to move education reform in a certain direction. Contrary to the principles of free market competition preached by international financial institutions, there is no competition when it comes to designing and coordinating education sector development in Mongolia. ADB has a firm grip on the Ministry of Education, and senior government officials are too intimidated to negotiate with other large donors for fear of undermining ADB's leading role in education reform, and upsetting their main financial ally. This monogamous bond between a government and an international donor is neither healthy nor inclusive. The gentlemen's agreement—or, more accurately, the business agreement—between ADB and the World Bank to divide up the sectors in which each of the donors takes on a leading role in Mongolia has had a diminishing effect on accountability. ADB safeguards its monopoly by limiting access to project documentation and evaluations and thereby hinders an informed and critical dialogue on education sector development in Mongolia.

The role of the World Bank in Mongolia is very much restricted to the cross-sectoral Poverty Reduction Support Fund, only marginally benefiting the educational sector (see chapter 9), and to analytical and advisory activities (World Bank 2004a: appendix 7).

Bilateral Donors

Whereas the donor logic of the Asian Development Bank and the World Bank is finance-driven, the logic of bilateral aid agencies is self-referential in a different way. The bilateral agencies of the German and Danish governments—Gesellschaft für technische Zusammenarbeit (GTZ) and Danish International Development Assistance (DANIDA)—selectively export "best practices" from their own educational systems that are supposedly missing or underrepresented in Mongolia. For example, German consultants have felt compelled—not only in Mongolia but also in many other countries—to contribute to vocational education. Meanwhile Danish experts focus on small schools and students with special needs. Once the Americans get involved under the auspices of the Millennium Development Fund, their specialists will emphasize English language and Information and Communication Technology reform in Mongolia. The decision of what to support in the Mongolian educational sector is more driven by what the lender has to offer than what the borrower actually needs. Over twenty years ago Brian Holmes (1981) astutely observed how country-specific preferences determine policy export. Writing during the era of the Cold War, Holmes found that, regardless of the circumstances, British and American experts almost always favored the introduction of a decentralized system of educational administration, whereas Soviet and German Democratic Republic experts always recommend the introduction of polytechnical education in the countries they advised.

This is not to downplay the importance of bilateral grants for educational development in Mongolia, rather it is a commentary on the reforms each bilateral cooperation agency tends to advance, and those which it neglects. No doubt, rural schools, for example, would be in much worse condition today without two large grants from DANIDA, amounting to approximately $9 million. From 1992 to 1998, the Danish government supported primary and secondary education reform in schools outside of Ulaanbaatar, and since 2000 has financed development in 80 rural schools (Steiner-Khamsi, Stolpe, and Gerelmaa 2004a). DANIDA's involvement in other areas such as nonformal education, life-skills training, and instruction for children with special needs and disabilities is less well known; it is, nevertheless, substantial.

Asian donors, in particular the Japanese International Cooperation Agency (JICA) and the Korea International Cooperation Agency (KOICA) are infrastructure and resource oriented ("hard-type" aid), but have not yet become involved into "soft-type" aid, such as the reform of content or methods in education. Both have shipped technical equipment for radio, television, and video-conferencing studios, as well as computers (some of which are secondhand) to educational institutions in Mongolia. Without much fanfare the Japanese government consistently channels grants into the building of schools. Between 2000 and 2002 alone, JICA reconstructed at the inflated cost of $24 million a total of 16 schools in Ulaanbaatar, and approximately the same number of schools in the towns Darkhan and Erdenet (Steiner-Khamsi and Nguyen 2001: appendix 2).

THE UN SYSTEM

It is common to merge United Nations organizations (UNDP, UNESCO, UNICEF, UNIFEM, UNFPA) with other intergovernmental organizations (World Bank, IMF) into the same category of multilateral donors. We refrain from doing so for two reasons. First, UN organizations have come to see themselves, in light of dwindling funds, as facilitators rather than as funding sources for international cooperation projects. The United Nations Development Programme (UNDP), the largest UN organization in Mongolia, for example, planned projects for the period 2002–2006 in the amount of $29.1 million. From these total costs only one-fifth is financed from UNDP regular resources, the remaining $23.6 million are mobilized from third parties and other funds (UNDP 2001: 11). Their budget is tiny compared to other multilateral donors. Second, as opposed to the banks, which only collaborated with market economies, the UN organizations bridged the internationalist (Second World) and international (First and Third World) space in transnational cooperation. This applies especially to UNESCO and UNDP, particularly in the period after 1954 when the Soviet Union rejoined the organization, and paved the way for other socialist countries, including Mongolia, to be admitted as member states. While there exist fascinating accounts of how UNESCO has transformed itself from a Western organization with a vision of "embedded liberalism" (Mundy 1999: 28) into an organization that gave voice to its numerous members from the Third World (Jones 1988; Mundy 1999), not much has been written about its metamorphosis from a Western Cold War institution to one that became truly universal. During the period of the Cold War, UNESCO and UNDP were among the few, if not the only, multilateral organizations with an international reach embracing both world-systems, that is, countries from market as well as from planned economies.

UNESCO grew from 58 member states in the period 1945–1973 to 153 members in the second period 1974–1984 (Mundy 1999: 29), and currently comprises 191 member states. The boom between the first and second period had two different causes: former colonies that gained independence in the 1960s and 1970s joined UNESCO, and socialist countries, after numerous failed attempts, were admitted as member states. The second increase mirrors the dissolution of the Soviet Union in the early 1990s that produced nearly 30 independent states, all of which became members of the UN system. What is missing from historical accounts of UNESCO, however, is an acknowledgment that several of the Third World countries which launched the 1974 UNESCO "revolution," and demanded a new international economic order were in fact socialist member states. The revolution was prompted by the withdrawal of the United States, Great Britain, and Singapore which, in turn, enabled UNESCO to further shift its emphasis toward Third World countries, and to a lesser extent, countries of the Second World.

Being admitted as a socialist country to the United Nations was not an easy endeavor. The Mongolian People's Republic tried for 15 years, and was only accepted as a member state in 1961. In this case, stubborn rejection was not only the response of Western market economies, but it was also that of the Eastern neighbor, China. Admission to UNESCO and UNDP was an important priority for the socialist government of Mongolia, and thus portrayed as a victory:

> Ever since 1961, when the socialist and other progressive states helped Mongolia take its legitimate place at the United Nations, our international prestige has been growing from year to year. (Luvsanchultem 1981: 137)

It is also important to bear in mind that both world-systems—the United States and its allies, and the Soviet Union and its allies—claimed ownership over the project of world peace, international understanding, and universal collaboration, inscribed in all UN institutions. The following illustrates the request from 1981 that Mongolia be treated as an "equal and active member of the United Nations" in the common cause for international peace:

> Our country [the Mongolian People's Republic] was one of the first to give its full support to the ideals and tasks of the United Nations. This is quite natural since, being a socialist state, it had, long before the foundation of the UN, been devoting its foreign policy to the task of consolidating world peace and to the development of co-operation among nations on the basis of the principle of the peaceful coexistence of states with different social systems. (Dugersuren 1981: 141)

The socialist government neither wished to be nor was, within its own world-system, isolated; on the contrary, it was involved in numerous internationalist cooperation projects. For example, the socialist government published a bimonthly magazine in English, obviously written for like-minded supporters in the capitalist world, with a section entitled, "Visit of Friendship and Brotherhood." Here the regular visits of leaders from newly emerging socialist movements or states, visiting Mongolia to learn and borrow from the flagship, low-income state socialist country, were recounted. The visit of the Frelimo leader from Mozambique, Samara Moises Machel, in 1978 was one such occasion, during which collaboration between MPRP and Frelimo in the "interest of international peace and internal security" were discussed (Mongolian People's Republic Council of Ministers and State Committee for Information, Radio and Television 1978: 5).

An analysis of the contemporary practices of UN organizations requires a distinction between the two UN players in Mongolian education: UNICEF and UNESCO. Phillip Jones has scrutinized the donor logic of multilaterals (Jones 1998, 2004), and found great differences, depending upon how they are funded. He points out that UNICEF relies on voluntary donations from governments, private foundations, or individuals, and therefore "its analyses of need tend to be dramatic, its projections tend to be alarmist and its solutions tend to be populist" (Jones 1998: 151). In contrast, UNESCO runs on membership fees that are, unfortunately, more successfully extracted from low-income governments than they are from high-income governments. Given the global scope of UNESCO's operation, supported by minimal funding, little ends up left at the country-level.

In Mongolia UNESCO projects do indeed run on a very low budget and depend heavily on mobilizing other donors. Precisely because of its good reputation in Mongolia, some projects are erroneously attributed to UNESCO even though they are funded by others. For example, the Gobi Women Initiative, the nonformal education project, and the life-skills project are funded by DANIDA and other donors, even though they are commonly associated with UNESCO. A cause for concern is that UNESCO is charged with coordinating the Education for All report on a very limited budget; this is particularly relevant given the great weight attached to the internationally agreed upon templates, which determine how an inventory of the state of education in a given country must be made. Jones's assessment of UNICEF's donor logic appears equally accurate in the Mongolian context. The byproduct of a marketing strategy that targets voluntary donations worldwide, the UNICEF-funded projects in Mongolia make an appeal to compassion and pity with little regard for larger developments in the education sector. A good case in point is UNICEF's initiative to have senior-level government officials and international visitors stay overnight in a rural boarding school so that they can personally experience the horrendous poverty in rural Mongolia.[12]

UN organizations in general, and UNESCO and UNICEF in particular, have adopted a distinctly convincing and commendable human rights approach to education. Their objectives are framed in terms of rights, especially the rights of women and children. Similar to other international organizations that backup their agendas with indicators, benchmarks, and monitoring instruments to ensure that their agenda is followed, UNICEF has developed a set of instruments to safeguard the rights of women and children worldwide.

UNICEF and UNESCO transplant modules, tool kits, checklists, reform packages, and "best practices" to different corners of the world. These activities are as extensive as those carried out by international financial institutions, and given more priority than they are with bilateral donors. UNICEF's Child-Friendly School (CFS) program is one of countless modules or packages that has traveled around the globe, and also ended up in Mongolia. Child-Friendly Schools—*Gud Fella Frenly Skul* (Bislama), *khüükkhed eeltei surguul'* (Mongolian)—are to be found wherever UNICEF is operating. The project is generally meant to enhance the quality of education, and the focus in Mongolia is on improving the learning environment. Other countries participating in the project in UNICEF's East Asia and Pacific region (to which Mongolia is assigned) are the People's Republic of China, Philippines, Thailand, and Vanuatu (UNICEF 2005). Priorities vary slightly in other countries, with different emphases placed on girls' education, life skills, or education in emergencies, but the marketing of UNICEF-funded schools as "rights-based" and "child-friendly" has been a key feature of its global campaign.

Nongovernmental Organizations

In 2005, there exist close to 4,500 associations that are registered as non-governmental organizations in Mongolia (Beck 2005). These NGOs, local and international, have grown exponentially since 1996, when the first legal foundation for a registration was created. Already by 1999 (UNDP 2000) 1,614 associations had registered as NGOs, and by the year 2003, 3,200 NGOs (Ariunaa 2003) had been established, operating predominantly in the areas of sports and recreation, the fine arts, and other cultural domains. Therefore from 1999 to 2003—only four years—the number of NGOs doubled. Of all the NGOs registered as tax-exempt entities with the Ministry of Justice, less than 5 percent offer services or programs in education, and many NGOs either exist only on paper, or consist of only 1 or 2 staff members who have sought a way to evade taxes. Since there is no monolithic definition of what constitutes an NGO other than the requirement that it serve either "society" or its members, there is tremendous terminological confusion on this issue. For example, at a conference in November 2003, organized by the Mongolian Foundation for Open Society, the question was raised as to why private hospitals and schools (charging fees) are not regarded as NGOs, given the positive nature of their work and the valuable services they provide to the community.

Similar to other postsocialist countries, where the legal status of nongovernmental organizations has only been established over the course of the past decades, the category NGO encompasses a wide range of associations. These range from GONGOs (government-organized NGO) or QUANGOs (quasi-nongovernmental organizations) to DONGOs (donor-organized nongovernmental organizations), and even private, tax-exempt businesses (see Fisher 1997; Edwards and Hulme 1996; CICE 1998). Ruth Mandel's identification of NGOs in Kazakhstan, Krygyzstan, and Uzbekistan as donor-organized NGOs (DONGO) also applies to a great number of NGOs registered in the field of education in Mongolia (Mandel 2002). She notes that many local NGOs in these countries master "NGO-speak" and focus exclusively on initiatives for which they are likely to obtain international funding, notably from bilaterals such as USAID, or from international NGOs such as the Soros Foundations Network, Save the Children UK, and so on. Once they are in the inner circle of DONDOs, their survival is secured, or as Mandel describes it,

> the simple fact that they have already received money from other Western donors serves as a recommendation to other potential donors. (Mandel 2002: 285)

UNDP Mongolia commissioned a study on NGO-implemented assistance in the social sector (UNDP 2000). The study surveyed a sample of 90 Mongolian and 10 international NGOs, and found that 93 percent of the funding of Mongolian NGOs is provided by international donors. This result confirms the prevalence of donor-organized NGOs, also found in other post-socialist countries. Their association with civil society—the idea that local NGOs are "closer to the people" (Edwards and Hulme 1996: 963), and therefore more responsive to local needs—is a myth shattered once it is recognized that a great number of NGOs function predominately as prolonged arms of international donors. There are, however, numerous Mongolian NGOs that receive very little press coverage or public attention, but serve populations that had been neglected either by the government or by international donors. These small, low-budget NGOs run schools for dropouts, provide family support, establish shelters for street children, and facilitate schooling for children with physical disabilities.

There are several large international NGOs in Mongolia such as the Mongolian Foundation for Open Society (MFOS), Save the Children UK, and World Vision that function not only as implementers of projects, but also as donors or grant providers. The educational programs of MFOS, for example, funded school and teacher education reform projects in an amount close to $3 million between 1998 and 2004. The donor logic of these international NGOs is very much vision-driven, in that they fund projects in Mongolia that correspond with the overall mission of the organization. Functioning as field-offices for headquarters that are based in New York (MFOS), London (Save the Children UK), or Canberra (World Vision), their operational budget and activity plans very much depend on decisions made abroad. This also applies to the Soros Foundations Network even though it purposefully labels its field-offices, like the MFOS, as "national foundations" and has national boards in place to ensure that national rather than international agendas are pursued. Nevertheless in 2004, the highly visible educational programs of the MFOS were transformed into a local NGO (Mongolian Education Alliance) because of changed priorities at the headquarters of the larger Soros Foundations Network. As a result of the shift from sector-specific projects to general public policy and advocacy work, most educational programs in the postsocialist region of the Soros Foundations Network were either terminated, or are currently in the process of being phased out. In Mongolia, what remained of the MFOS was transformed into an NGO (Open Society Forum) that functions as an effective and outspoken policy think tank as well as a public resource and media center that advocates greater civic involvement and transparency in public policy.

Several scholars (Mundy and Murphy 2001; Appadurai 2000) have referred to international NGOs as "transnational advocacy networks" (TAN) or treat them as global social movements (Boli and Thomas 1999: 47) that—in their capacity as a global civil society—influence not only world politics but also national governments. Established for the purpose of "doing good" (Fisher 1997), international NGOs usually adhere to values of social justice, civic participation ("civil-society-building"), and transparency in education. Arjun Appadurai (2000) also had the good cause of NGOs in mind, when he proposed that TANs should be viewed as the only international force to counter corporate globalization. Economic globalization is, according to Appadurai (2000: 16), "a runaway horse without a rider" which can only be tamed by successful TANs that "might offset the most volatile effects of runaway capital." If NGOs become united among themselves, they would have the right values, the organizational structure, and the universal reach to steer a "globalization from below" which would bring social justice and light to the darker sides of (economic) globalization. Judging from the context for NGOs in Mongolia, this is unlikely to happen. NGOs, or more accurately the DONGOs in Mongolia, compete among themselves for the scarce resources of international donors, and they do not see themselves as part of a global movement, but rather as local counterparts for headquarters that are based in the capitals of corporate globalization.

Arguably, there is more choice and diversity if one compares international donors with the framework of internationalist cooperation during the socialist period. Internationalist technical assistance meant the transfer of only one, Soviet reform package to Mongolia, whereas international donors nowadays each promote and fund their own package or "best practices." While dependency on external assistance has remained constant in Mongolia, the allocation of external funds to the social sector—as is be presented in further detail in the next chapter—has dramatically decreased over the past 15 years.

5

Structural Adjustment Reforms, Ten Years Later

> The nation of Mongolia is in the process of a difficult and often frustrating transition to a democratic and free market society. The structural adjustment necessary for this transition can have disproportionate impacts on the education and human resource (EHR) sector if the sector's key role in the transition is not carefully articulated. Serious damage to the present education and training system could even cause a decline in public acceptance of the structural adjustment process itself. (Government of Mongolia 1993: i)

These introductory lines of the 1993 Mongolia Sector Review mark a historic moment: the submission to structural adjustment reforms in education. Compared to other sectors, the educational sector in Mongolia was relatively slow in giving way to international pressure to "adjust," or more accurately, to downsize educational provisions. In education, rampant structural adjustment reforms were carried out in 1997 and 1998, and these reforms were met with vociferous public protests. In education, the adjustments were commonly regarded as externally imposed, and therefore encountered no "public acceptance of the structural adjustment process" (Government of Mongolia 1993: i). In more of a *preview* than an actual *review*, the 1993 Mongolia Sector Review outlined how the educational sector would be restructured, as well as how it would be aligned with the overall structural adjustment policies (SAPs) of the 1990s.

International financial institutions, notably the International Monetary Fund (IMF), but also the World Bank and regional banks, impose structural adjustment packages if they conclude that a low-income country is living "beyond its means" (Ray 1998: 690). The rationale is straightforward: The international financial institutions are, first and foremost, banks; they only give loans under the condition that low-income governments introduce "appropriate policies" (IMF 2004: 1) to increase the chances of loan repayment. The mandate of IMF is to advance global trade, foreign investments, and the transnational flow of money regardless of the impact it makes on national economies. Banks also include long-term economic development and growth as desired goals and, since the 1990s when their structural adjustment projects came under serious attack, added poverty-reduction and good governance as conditional terms for loans.[1] SAPs are comprehensive reforms that force governments to liberalize prices, devalue their currency, and reduce public expenditures. In postsocialist countries, privatization has been added to the global SAPs agenda. In practice, SAPs have led to massive unemployment, poverty, and huge external debts (Stiglitz 2003), and have been accused of only benefiting elites in low-income, and entrepreneurs in high-income countries. In the past decade, two additional features of SAPs have been frequently criticized: their imposition and their velocity.

First, the fact that governments sign off on SAPs, sector strategies, and sector reviews such as the one cited above, tends to conceal the reality that these agreements are not only funded but also frequently designed and imposed by international financial institutions. In an attempt to increase government ownership, the sector-wide approaches (SWAp) emerged in the second half of the 1990s to encourage governments to formulate a development strategy, and then request international donors to contribute financially (see Appadu and Frederic 2003; CICE 2001). With the exception of the Poverty Alleviation Program (1994–2000) in which the World Bank, UNDP, SIDA, ADRA, the Government of the Netherlands, and 10 other international donors contributed a total of $19 million (Government of Mongolia, World Bank, and UNDP 1999: Table 5), we have not yet come across another SWAp in the Mongolian social sector. In stark contrast to the Poverty Alleviation Program (with admittedly limited impact),[2] the Mongolian education sector development reviews and programs have been single-handedly designed, funded, and executed by the Asian Development Bank (ADB).[3] The unilateral approach of ADB is evident throughout various stages of a program. Government officials and Mongolian experts are used as informants at the initial stages of needs assessment and project appraisal, but have little to say on the final versions of the project documents. The programs are implemented by a separate ADB-entity hosted in the Ministry of Education (Program Implementation Unit) that is both accountable to ADB and the Ministry of Education.

Second, the pace with which the SAPs have been implemented was so traumatic for the postsocialist economies, and for the residents living in them, that leading development economists such as Padma Desai wonder why a more "gradualist" approach was not taken into consideration at the time (Desai 2002: 223; see also Desai 1997). Poland was the first to undergo "shock therapy," a method of fundamental economic reform that was subsequently transplanted to the remaining countries of the postsocialist world. Under the tutelage of IMF, Jeffrey Sachs designed a fast-paced structural adjustment strategy that aimed to privatize all state property, boost economic activity by reducing state intervention, devalue the currency, and liberalize prices within a mere 400 days (see Nolan 1995: 268). The shock therapy approach, also referred to as Poland's "big bang" (Sachs 2005: 123), was applied next to the Russian Federation. The "idea of the Grand Bargain" in Russia is explained by Sachs as follows: In return for accelerating economic reforms and democratization, the government receives "large-scale financial assistance" from the United States and Europe (Sachs 2005: 133). Reflecting on the impact of shock therapy on the residents in Russia, Sachs writes,

> Many critics later accused me of peddling a ruthless form of free-market ideology in Russia. That was not the case. My main activity for two years was an unsuccessful attempt to mobilize international assistance to help cushion the inevitable hardships that would accompany Russia's attempt to overcome the Soviet legacy. (Sachs 2005: 137)

In Russia, shock therapy began on January 1, 1991, and the postsocialist government in Mongolia followed in the footsteps of its "older brother" *(akh)* a few months later. In Mongolia, the privatization of livestock and agricultural land went into effect in September 1991. After the first year of privatization, only 57 of the 255 animal husbandry collectives (*negdel*) survived in the form of newly established cooperatives; 40 were completely disbanded, and the remaining 158 *negdels* generated 320 privately owned companies (Korsun and Murrell 1995).

The first two years of the economic transition (1991–1993) were a period of economic collapse, pressing the country to the verge of a hunger crisis. The removal of stable prices for consumer goods had a disastrous effect on inflation. As a result of price liberalization, gasoline prices increased fourfold in 1991 (World Bank 1992: 19), and the overall inflation rate exceeded 400 percent in 1992 (Government of Mongolia 1993: section 1–28). Unemployment grew exponentially during the same period, and the number of residents who could not live off their salaries or pension plans became sizeable. However, a full account of Mongolia's economic trauma must also consider the broader context: in the early 1990s when shock therapy was introduced, an economic collapse was already imminent. Upon closer examination, it becomes clear that Mongolia was actually exposed to two economic shocks in short succession.

The first shock concerned the economic hardship in other former socialist countries and the collapse of CMEA. The collapse of the Soviet Union (1989) was particularly felt in Mongolia, when CMEA's assistance to Mongolia (1962–1991) dissolved. Additionally, the withdrawal of assistance and the suspension of trade with the Soviet Union and other socialist countries (accounting for over 90 percent of Mongolia's exports) on January 1, 1991, led to major economic decline. According to Peter Boone (1994: 330), in 1989 internationalist assistance amounted to 53 percent of Mongolia's GDP, and in 1991 it dropped to 7 percent of the GDP (gross domestic product).[4] The second economic shock was caused by Mongolia's admission to the International Monetary Fund, the World Bank, and the Asian Development Bank in February 1991 (Batbayar 2003). As conditions for being admitted to this new sphere of external assistance, the government had to import a structural reform package and rigorously implement it in all sectors, including the education sector.

Educating "Beyond Their Means"?

Strikingly, each and every public expenditure review uncritically recycles the harsh attack on the Ministry of Education for spending too much on education (World Bank 1992; Government of Mongolia 1993; Government of Mongolia and ADB 2001; Bartlett, Byambatsogt, and Enkh-amgalan 2004). The reports of international finance experts include statements such as the following:

> *Mongolia spends more on education than most low-income countries* [italics in original]. Total spending on education was 5.7 percent of GDP in 1997, compared to 3.5 percent in the East Asian and Pacific region overall, 3.7 percent in all low-income countries, and 5.0 percent in OECD countries. (World Bank 2002: 127)

For the past decade, the international financial institutions in Mongolia powered the generic structural adjustment formula (generating income and reducing expenditures) as summarized in the first Education Sector Review of 1993: "Generation of additional resources should be accompanied by efforts to reduce costs" (Government of Mongolia 1993: section 2–20). To date, this two-pronged formulate occupies a position of sacrosanct priority to international financial institutions operating in Mongolia. It casts a shadow over all other shortcomings of the educational system that deserve far greater attention, notably the rapid decline of universal access to education.

In 2000, the government allocated 7 percent of the GDP to education (Bartlett, Byambatsogt and, Enkh-amgalan 2004: 17). The accusation of overspending for education stands and falls with the validity of the comparative framework (how high is the allocation in Mongolia when compared to other countries?), and the reliability of the percentage (how high is the GDP in reality?). Arguably, 7 percent of public expenditures for education as a percentage of GDP appears high *if* compared to the "region," and *if* GDP is taken at face value. However, such an assessment of overspending in education is, as the following attempts to illustrate, based on two faulty premises.

Comparison with Countries from the Wrong Region

First, Mongolia is placed in the wrong region, and the validity of comparative statements is therefore seriously hampered. For the sake of easier management, Mongolia has been erroneously placed in one of the following two regions: (1) the World Bank, and the UN organizations UNDP, UNICEF, and UNESCO categorize Mongolia as part of the Asia and Pacific region, and (2) the Asian Development Bank counts Mongolia as part of the East and Central Asia region. This means that educational development in Mongolia is always compared either with

> Cambodia, China, Cook Islands, Fiji, Indonesia, Kiribati, Democratic People's Republic of Korea, Lao People's Democratic Republic, Malaysia, Marshall Islands, Federal States of Micronesia, Myanmar, Nauru, Niue, Palau, Papua

New Guinea, Philippines, Samoa, Solomon Islands, Thailand, Timor-Leste, Tokelau, Tonga, Tuvalu, Vanuatu, Viet Nam. (World Bank 2005 and UNICEF 2005)

or with

Azerbaijan, People's Republic of China, Hong Kong/China, Kazakhstan, Republic of Korea, Kyrgyz Republic, Taipei/China, Tajikistan, Turkmenistan, Uzbekistan. (ADB 2005)

A landlocked country with a strong orientation toward countries in the West and the North, misclassification implies that Mongolia is included in regional initiatives that are only of relevance for island states or states bordering an ocean. Most international NGOs, with the notable exception of the Open Society Institute, also misclassify Mongolia as part of the Asian and Pacific region.[5] Interestingly, the World Bank set up its regional categorizations in a way that treats Europe and Central Asian countries as a separate region (ECA region). Similarly, the UN system counts countries from Central and Eastern Europe (CEE), the Commonwealth of Independent States (CIS), and the Baltic states as part of the same cluster. But Mongolia is conspicuously missing from the World Bank's ECA region and the UN's CEE/CIS/Baltic region, ones that cover former socialist countries.

When it comes to Mongolia, the artificial creation of a space is so misguided that one wonders what global power is assumed in redesigning geopolitical maps at the will of international organizations. This positioning of Mongolia in the wrong region neglects the history of Mongolia, notably its cultural proximity to Central and Eastern Europe, the Caucasus, the Soviet Union, and former Soviet republics in Central Asia. The remapping exercise of the Asian Development Bank is slightly more precise in that it also includes comparable cultural contexts such as Kazakhstan, Kyrgyzstan, Tajikistan, Turkmenistan, and Uzbekistan. However, only in 2005, eight years after ADB established the Central Asia Regional Economic Cooperation program, was Mongolia admitted to join the group of Central Asian countries participating in the regional initiative. One would be hard-pressed to acknowledge that this was due to a higher degree of cultural sensibility. More likely, it had to do with the fact that the Asian Development Bank exclusively operates in Asia, and assigns countries to subregions that make their operations more manageable. Manageability as a criterion for redrawing geopolitical maps, naïve as it is, has had a major negative impact on educational development in Mongolia. A different range of educational spending emerges if international organizations stop comparing Mongolia to noncomparable educational systems in East Asia and the Pacific that supposedly cohabit the same "region," and choose appropriate units of comparison instead.

In the following we present two maps. The first map (map 5.1) illustrates where the UN system currently positions Mongolia (Asia and Pacific region);

Map 5.1 Map of Mongolia's Region as Defined by the UN System
Source: UNICEF 2005.

Map 5.2 Map of the Former Socialist Region from which Mongolia is Excluded
Source: UNDP 2005. Note that all former socialist countries except Mongolia are highlighted as part of the same region.

the second map (map 5.2) shows how Mongolia is left out of the UN region in which it would be more appropriately placed (ECA region or CEE/CIS/Baltics).

Once we compare educational systems that are in fact comparable, it becomes clear that Mongolia does not spend that much more on education than its counterparts. In 1995, it spent 6.3 percent of the GDP in the education sector, a lower percentage than what governments in Estonia (7.5 percent), Georgia (7.7 percent), Kyrgyzstan (6.6 percent), Latvia (6.6 percent), Moldova Republic (9.0 percent), and Ukraine (7.2 percent) spent (Bartlett, Byambatsogt, and Enkh-amgalan 2004: 17) on education. The following Figure 5.1 illustrates that it was only in 2000 that the Government of Mongolia visibly outspent all other former socialist countries.[6]

Clearly, educational spending dramatically dropped in all postsocialist countries in the early 1990s. This applies whether we use the percentage of GDP that is allocated to education as a measure (see Figure 5.1), or educational spending as a percentage of public expenditures (UNICEF 1999: 5). The picture of rapidly falling public expenditures for education is especially grim in the Republic of Georgia, where educational spending previously accounted for almost 36 percent of all government spending but, within only 4 years (1993–1997), was slashed by more than half. In 1997, the Government of Georgia was only able to commit less than 15 percent of its public spending to education (UNICEF 1999: 5). Despite the pressure to significantly curb educational spending, several governments succeeded in increasing the percentage of GDP allocated to education by the end of the

Figure 5.1 Percentage of GDP Allocated to Education (Transition Economies)

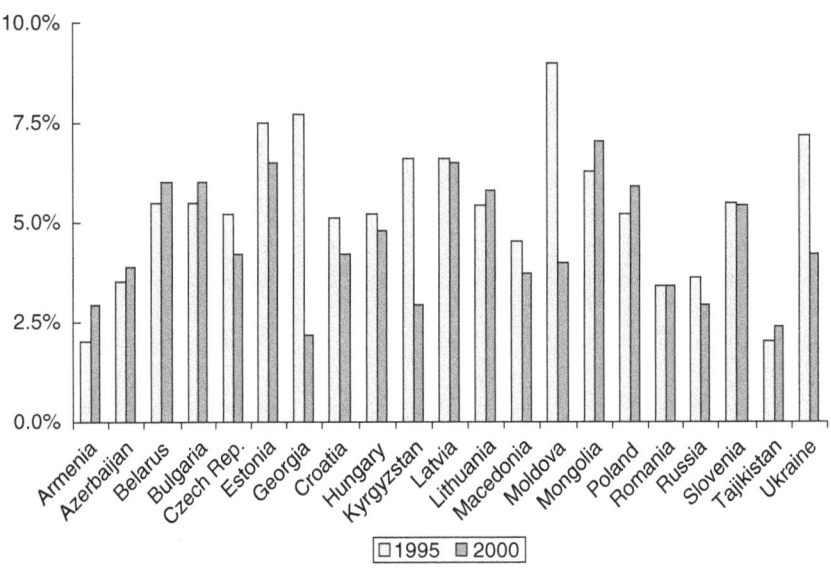

Source: Bartlett, Byambatsogt, and Enkh-amgalan 2004: 17.

decade (Armenia, Azerbaijan, Belarus, Bulgaria, Lithuania, Mongolia, Poland, and Tajikistan).

In Mongolia, approximately one-fifth of all government spending is earmarked for education. In light of developments in the region, this strong commitment to education is noteworthy. This means that the question of "overspending" should be more accurately reframed: Why has the Government of Mongolia, despite the structural adjustment pressure, preserved a high level of financial commitment to the education sector?[7] The most common explanations, which are also acknowledged by international donors, are the sheer number and the kind of students to which the sector caters, as well as the weather conditions with which the sector has to cope. At the end of the 1990s, slow population growth played a crucial role in tipping the balance between adults and children in Mongolia. Until then, children under 19 years of age represented the majority of the total population (UNICEF 2000: 7). To this day, the number of school-aged children remains large compared to other educational systems. Additionally, in order to provide access to formal schooling for a student body that is both widely dispersed and nomadic, the sector has to maintain an expensive system of boarding schools. Today, these residential schools are only able to admit 5.5 percent of the total student population,[8] although many more children from poor herder families depend on having a dormitory space in order to attend school. Finally, it is also uncontested that the harsh climate requiring a long heating period (October 15–April 15) furthers the justification for high public expenditures in education. This is not to suggest that the entire budget allocated to education is necessarily spent wisely or efficiently;[9] however, we argue that it would be a fallacy to compare the Mongolian education system with other systems that deal with a smaller and dissimilar student population, different environmental conditions, and last but not least, with a lower level of public commitment to education.

Comparison with the Wrong Amount of National Income
Second, the accuracy of GDP calculation must be questioned. More specifically, does the proportion of GDP spent on education still hold when we consider the informal economy, the shadow economy, the underground economy, or the parallel economy? There are many reasons to believe that the GDP in Mongolia is much higher than officially acknowledged. As a corollary, the proportion of public expenditures on education as a percentage of the complete GDP—formal and informal economy—is much lower than 7 percent.

Several studies have dealt specifically with the shadow economy in Mongolia (Anderson 1998; Bikeles, Khurelbaatar, and Schelzig 2000; Amar 2004), and surveyed small shopkeepers, small manufacturing enterprises, kiosks, cabdrivers, informal currency dealers, street vendors, and so on who all have one thing in common: they do not report their income, do not pay taxes, and their income is not included in the official GDP calculation. In 2004, for example, N. Amar used an index utilized by the Organization for

Economic Cooperation and Development (OECD) to estimate the size of the shadow economy in Mongolia. Based on rough estimates of the informal economy in other transitioning and developing countries, Amar (2004: 34) concluded that the size of the Mongolian shadow economy was 32.5 percent of the gross domestic product (GDP) or $335.7 million, respectively, in the year 2002. This means that the Government of Mongolia loses over $300 million in tax revenues from unreported economic activities, an amount not included in the official figure of the GDP. The shadow economy in Mongolia is large, but if seen from a comparative perspective it is in the middle range compared to other postsocialist countries. The country with the biggest parallel economy is Georgia (64 percent of the GDP). Russia's informal economy is 44 percent of the GDP, and Uzbekistan's shadow economy comprises only 9 percent of the GDP. These data need to be read cautiously as semi-illegal and tax-exempt undertakings, informal economic activities are difficult to define and measure. For example, the *2004 Human development report* (Government of Mongolia and UNDP 2004: 41ff.), devotes a section to the informal sector in Mongolia, and lists several studies that attempt to measure the scope of this sector. The 1997 survey by the World Bank and the 1999 survey by USAID came up with much more conservative figures to capture the size of the informal economy—under 15 percent as a percentage of the official figure of GDP—than Amar estimated (2004). Why each study's figures on the informal economy vary so drastically is a subject that deserves further scrutiny.[10] Despite the methodological difficulties with quantifying economic activity in the informal sector, it is uncontested that an exclusive reliance on the formal economy distorts real income figures in Mongolia.

On the positive side, the well-developed informal economy keeps many households from slipping below the poverty line. Additionally, it is important to bear in mind that the informal economy is not a "transition" phenomenon in Mongolia, but rather it is an established tradition. Several scholars have reconstructed the role of informal economies during socialist times, dismantling the assumption that all financial transactions were completely controlled by the "command economy." Besides the permission to accumulate private possessions, albeit to a limited extent,[11] residents engaged in all kinds of informal economic activities to purchase desired commodities, and relied on their personal and political networks to get things done in daily life. Alena Ledeneva (1998: 1) views the Soviet practice of using personal networks (Russian: *blat*) to "obtain goods and services in short supply and to find a way around formal procedures" a pillar of socialist systems. She considers the informal economy and these exchange networks in particular

> the "reverse side" of an overcontrolling centre, a reaction of ordinary people to the structural constraints of the socialist system of distribution—a series of practices which enabled the Soviet system to function and made it tolerable, but also subverted it. (Ledeneva 1998: 3)

The argument advanced by Ledeneva is that these decentralized, unplanned, and informal economic transactions were vital for survival in an economy of shortage, and in effect "contributed to the functioning of the Soviet system" (Ledeneva 1998: 3). The socialist informal economy has also sparked the interest of economists. For example, in his comparative analysis of informal economy activities in the 15 Soviet republics, Byung-Yeon Kim (2003) found that the average Soviet household expenditure in the informal economy (as a share of total household expenditure) was 23 percent in the period 1969–1990.

More Than Loose Coupling: Local Responses to Imposed Reforms

The pressure on the Government of Mongolia to reduce government spending in education is based on a figure that is seriously skewed. Despite misguided comparisons and unreliable data, international advisors insist, report after report, that Mongolia is living "above its means" when it comes to financing education. This chapter presents three examples of structural adjustment reforms in education that were imposed more than a decade ago: (1) tuition-based higher education, (2) decentralization of educational finance and governance, and (3) rationalization of staff and reorganization of schools.

In policy studies, it is common to differentiate between policy talk, policy action, and policy implementation (Tyack and Cuban 1995), or, more generally, to assume loose coupling between envisioned and enacted reforms. As conditionality for loans, the structural adjustment policies were not so much envisioned by Mongolian government officials as they were imposed by international financial institutions. This is not to suggest that Mongolian stakeholders in education—government officials, teachers, and parents—are merely helpless victims of imposed reforms. Since educational development means various things to different stakeholders in education, government officials were, more than once, forced under public pressure to cushion, offset, or withdraw earlier structural adjustment reforms. Thus, what comes across as "loose coupling," needs to be interpreted sometimes as active or subtle resistance.

We have selected these three structural adjustment reforms as examples of different policy encounters. Ten years later, one reform is still in place (tuition-based higher education), one moved back and forth (decentralization of educational finance and governance), and one partially reverted to a structure that was in place prior to the structural adjustment reforms (rationalization of staff and reorganization of schools). In this chapter, we draw in great part from three of our studies: a detailed analysis of policy documents and media reports on structural adjustment reforms (Stolpe 2001), a small qualitative study entitled Teachers as Parents (Steiner-Khamsi, Tümendemberel, and Steiner 2005), and an evaluation of 40 schools that are involved in the Rural School Development Project funded by DANIDA (Steiner-Khamsi, Stolpe, and Gerelmaa 2004a).

Tuition-Based Higher Education

Higher education was not only the first, but also the most fast-spaced and comprehensive reform of the early 1990s. During socialist times, roughly only 10 percent of an age cohort was admitted to higher education, and all advanced research was concentrated in the Institutes of the Academy of Sciences, reducing universities to professional training sites (Weidman, Yeager, Bat-Erdene et al. 1998). The reforms of the early 1990s completely revamped the socialist system of manpower planning in which the various ministries "owned" universities, and determined who as well as how many students were admitted to each field of study. The transformation from a vertical to a horizontal structure, whereby only one ministry (Ministry of Education) became the umbrella for all colleges and universities, was a major change to the former Soviet-type higher educational system (see Heyneman 2004: 4). Other reforms followed including the authorization to open private universities and charge tuition, or to replace the cohort-system with the credit-system. The idea was to deregulate the higher education sector, reduce state support and intervention, and attract private sector involvement. In contrast to the pre-1990s period in which each ministry had a large unit in charge of vocational and higher education, the Ministry of Education now hosts a consolidated higher education unit that is operated by only five staff members. The unit grants authorization to open private colleges and universities, checks on the facilities, and regulates accreditation. The only advantage of accreditation is eligibility for student grants and, the virtually nonexisting, student loans. The majority of privately owned colleges and universities do not bother assembling the piles of documents needed to apply for an accreditation. The demand for higher education is so high that most colleges and universities do not take on the bureaucratic hassle of accreditation.

These reforms opened up a huge market for private enterprises. In 1990, all colleges and universities were state-owned and free of charge. In 1992, a few private higher education institutions started to charge tuition, and a year later, the Ministry of Education mandated that all higher education institutions, private and public, collect tuition fees. By 2002 only 47 institutions were state-owned, 129 were privately owned, and 7 were run by international enterprises. Although private colleges and universities are much smaller in size than public ones, they serve a growing number of students. By academic year 1995–1996, every fifth student was enrolled in a private institution of higher education (8,930 students), and in academic year 2002–2003, every third student studied at a private college or university (37,607 students). It is important to mention that the Ministry of Education does not only actively support private sector involvement in education; it also ensures that public higher education is less attractive by imposing annual tuition that are approximately the same (amounting to eight monthly salaries of a teacher) as at private colleges and universities (Steiner-Khamsi and Nguyen 2001; see also Innes-Brown 2001). Advocates of structural adjustment celebrated higher education reform as a success, but viewed the reform as a disaster from a professional, social, and human angle. It created a host of problems that are heatedly discussed in the media, such as the overall drop in the quality of education in colleges and universities, the production of degrees that are worthless on the job market, unequal and corrupt admission procedures, fraud in awarding degrees and titles, and financial hardship for families.

The Teacher as Parents study (Steiner-Khamsi, Tümendemberel, and Steiner 2005) examined the financial hardship of parents, and in particular teachers, who have to pay for the education of their university-aged children. We wondered how exactly teachers managed to bridge the gap between high expenses and low income in education. The calculation on the income side is the following: A schoolteacher earns, on the average, T 639,400 ($570) in base salary,[12] and an additional 43 percent in bonuses and salary supplements a year. Thus, the full annual income of a teacher is on average T 925,800 ($824). On the educational expense side, the annual cost for tuition and living expenses of a university student ranges from T 681,200 to T 1,623,000 per year (Otgonjargal 2004: 25). This means that tuition and living expenses for one child absorb in the best case scenario approximately three-quarters of the full income of a teacher, and in the worst case twice the full income. Clearly, a teacher cannot afford to have her child enrolled in higher education unless she finds a creative way to narrow this gap.

We have identified three strategies that teachers in our study use to cope with the financial burden in higher education: reciprocity within the family, social redistribution within the extended social network, and loans.

Reciprocity Within the Family

From the 47 married female teachers in our sample, half of them (23 teachers) lived with an unemployed husband and therefore could not rely on additional income from their spouse. Financial support from retired parents (in the form of pensions) and support from children are very common. For teachers over 45 years of age, the financial support of their children was the most common type of family support. It deserves a longer explanation of how support of siblings works. Most frequently, the older children allocate a part of their monthly salary for the educational costs of their younger siblings. One of the recurring rationales given in our interviews was that their parents had made great financial sacrifices for granting them a "good education," and therefore they felt compelled to reciprocate by enabling their younger siblings to have the same opportunities. With the government restrictions for scholarships in the 1990s, only the oldest child became exempt from tuition fees. Since 2000, civil servants receive a scholarship in the amount of T 280,000 for one of their children. In return, the oldest children often take on the moral obligation of helping their parents pay for the educational expenses of younger siblings. In our interviews, we were told many stories of sacrifice among siblings. We confine our examples to the stories of three teachers at a rural district school in Övörkhangai:

24-Year-Old Mathematics Teacher (male, no children)
(Full monthly salary: T 60,000 composed of T 43,000 base salary, and T 17,000 bonus)

> I financially support my three younger sisters. Two of them are at university, and my youngest sister is in a class for mathematically gifted children at our school. I pay T 20,000 annually for her enrollment in that class, and an

additional T 10,000 for books and school supplies. Per month, I hand over approximately half of my salary to my mother so that she can pay for the education of my three younger sisters. I decided to get married in four years, when my two younger sisters will have completed their university degrees.

52-Year-Old Elementary School Teacher (female, 5 children; age range: 20–31 years old)
(Full monthly salary: T 60,000 composed of T 43,000 base salary, and T 17,000 bonus)

My first child lives in the same *sum* [rural district] and is unemployed, the second child works as a teacher in the same school and has two children, the third child studies at the Science and Technical University in Ulaanbaatar, the fourth child studies at a private university in Ulaanbaatar, and the fifth child graduated last year from tenth grade and then started to study at a public university in Ulaanbaatar. But during the academic year, she had to quit her studies, because I couldn't pay the tuition. I cannot afford having three children enrolled at the same time. Therefore we have decided to wait until one of the older two children completes the university degree. After that, my fifth child will re-enroll in university.

41-Year-Old Russian Language Teacher (female, 5 children; age range: 9–23 years old)
(Full monthly salary: T 78,000 composed of T 53,00 base salary, and T 25,000 bonus)

My second child used to study at the Science and Technology University in Ulaanbaatar, and I paid T 460,000 for tuition. In addition, I gave her T 300,000 for books and clothes. Her living expenses in Ulaanbaatar were very low, because we set up a small four-wall *ger* [Mongolian dwelling] in the *khashaa* [yard] of my older son in Ulaanbaatar, and she could eat and live there for free. However, this year she had to interrupt her studies because I needed all the tuition money to pay for the wedding of my older son. By next September, I will have enough money, and she can re-enroll in university. The education of children in secondary school is not cheap either, especially if you want to give them a good education. I pay approximately T 40,000 per year for books, clothes, and special events. The most expensive was the IQ club. I had to make my son quit the club because I couldn't afford the T 20,000 club fees.

The common thread throughout these three stories is that the professional, academic, and personal biographies of siblings are interconnected. Sisters and brothers make sacrifices for the sake of more important events occurring in the lives of their siblings: they interrupt studies or postpone marriage until their turn has come. This also means that parents keep all their children in mind when it comes to financially planning for higher education.

Social Redistribution Within the Extended Social Network

Another enigma emerged over the course of our data collection: Not 1 of the 44 interviewed teachers in our sample reported having borrowed money from friends and relatives, but a quarter of them pointed out that friends and relatives owed them money. The latter statement was typically accompanied with the complaint that this was a form of abuse, and that they do not anticipate their friends and relatives repaying their debts. Friendship ties and family bonds seem to suffer from broken financial agreements. This asymmetry deserves interpretation: A key to understanding this puzzle is the perception that richer relatives and friends have the moral obligation to help poorer relatives and friends. What they pass on monetarily is not considered a loan, but a gift that does not need to be returned. We have chosen to label this phenomenon "social redistribution" as it ensures that the rich members of that network do not become too rich, and that the poor are prevented from becoming too poor. This redistributive function is naturally resented by those situated in the higher echelons of income within the network.

Loans

One of the most striking findings of the Teacher as Parents study was that 42 percent of all respondents (27 teachers) are heavily indebted from a salary loan. In 1 of the rural district schools, each of the 13 interviewees reported having taken a salary loan, ranging from T 200,000 to T 900,000 ($178–$800). In the other three schools the finding was not as consistent, and only teachers with university-aged children were highly indebted. As an elementary school teacher in one of the rural district schools explained:

> I have so many expenses, and I don't know how to make ends meet. Sometimes I have a "loan depression" (*yadargaa*), because I need to keep borrowing money from friends and relatives in order to pay back my loans.

The loan interest rates range from 3 to 4 percent *per month*. The loans are either signed with the local bank or with the school. Many schools have started to function as banks and they provide salary loans at a comparable or sometimes lower interest rate than ordinary banks. Needless to say, teachers are bound to their workplace until they have repaid their loans. A creeping form of indentured labor, the gradual transformation of schools into banks is cause for alarm, but has not yet received the public attention in Mongolia that it deserves.

Teachers do not have to make greater sacrifices than the rest of the population to cover the high private costs of higher education for their children. On the contrary, as civil servants they are in a somewhat privileged position of receiving a government scholarship for one of their children.[13] However, teachers see it differently. In our interviews they conveyed a feeling of professional and personal tragedy. Professionally, they have experienced a sense of status loss over the past decade: the salary is stable, but low compared to others working in the private sector. Furthermore, their privileges as civil servants have been periodically cut, and now government scholarships for one child are at a bare minimum. Their responses to our interview questions reflected a combination of self-pity and dismay about the status loss of teachers in Mongolia. They tended to see themselves trapped in a double bind: As teachers they are viewed as the educated elite, and yet, they have difficulties ensuring that their own children receive the same level of education that they once received. Over and over, we heard the interviewees lament, "All my life, I sacrificed my life for the children of other people, and I don't have anything to offer to my own child."

A country of education-minded people, all parents in Mongolia wish to see their child enroll in higher education, even when they cannot afford it and in spite of the limited value of some higher education degrees. It would be mean to conclude that families have learned to cope with the financial burden that this early structural adjustment reform has bestowed upon them. Indeed they have succumbed, but at a high price of personal sacrifice. At the same time, the reform has contributed to social inequalities wherein children from poor families are locked out from the system of tuition-based higher education.

Decentralization of Educational Finance and Governance

The 1993 sector review recommends decentralization of governance and finance as a panacea for everything that purportedly went wrong during socialist times: the lack of quality, efficiency, and cost-effectiveness, and the dependence on external subsidies for funding the costly educational sector (Government of Mongolia 1993). From a broader international perspective, we concur with Joel Samoff (1999) that the education sector studies developed in the early 1990s are strikingly similar. The general crisis tone in Mongolian education sector studies when "diagnosing the problem," and the prescriptions for remedying "the problem," resemble, to a great extent, what Samoff (1999: 25) summarized for education sector analyses on the African continent. Samoff observed that the diagnostic section of the education sector studies often finds that the country is "in crisis," and that ". . . the government cannot cope, quality has deteriorated, funds are misallocated, management is poor and administration is inefficient" (Samoff 1999: 25). The prescriptive section of sector reviews, in turn, tends to recommend the following:

> reduce the central government's role in providing education; decentralize; increase school fees; encourage and assist private schools; reduce direct support to students, especially at tertiary level; introduce double shifts and multi-grade classrooms; assign high priority to instructional materials; favor in-service over pre-service teacher education. (Samoff 1999: 25)

In Mongolia all of these apply except for the "introduction of double shifts" (two existed, and a third was added in many urban and semi-urban schools) and "multigrade classrooms" that were attempted, but remained unpopular among parents and teachers.

What began in Mongolia as a presentation of possible solutions in 1993, albeit tainted by an international agenda that sought external assistance, was soon prescribed as a condition for international loans and grants. Decentralization of governance and finance was top on the list of conditionalities. A retrospective look at educational finance and management reforms reveals a curious pattern: For the past decade the Ministry of Education has periodically oscillated between decentralization and recentralization policies. On paper, it has consistently and enthusiastically subscribed to decentralization, but in practice has given these policies low priority (see Steiner-Khamsi and Stolpe 2004). The 2002 Education Law, locally referred to as the "recentralization law," drove a thick nail through the coffin of the decentralization policy. It was issued after a protracted period of international assistance: two large loans from the ADB and numerous projects funded by international organizations, over a period of ten years, to encourage the Ministry of Education decentralize the governance and finance of education. The 2002 Education Law was enacted at a time of relative ease, that is, seven months after a large loan from the ADB was secured.

The year 2002 was eventful for undoing the decentralization policies of the past decade. The Education Law downgraded the provincial Education and Culture Centers to provincial departments of the Ministry of Education. In 2002, they were renamed Education and Culture Departments, reporting since then to the Ministry of Education rather than to the provincial government. At the school level, the school councils were reduced to a consultative status and removed from all decision-making power with regard to finance and human resource management. The majority of the school council (9–11 members) had to be "founders" (*üüsgen baiguulagch*) of the school (MOECS 2002: section 35.1), that is, representatives of the Ministry of Education at the local level. In our study we found that there is a common understanding that the district governor and his officers represent the "founders," leaving at best two seats for parents, two for teachers, and one for students (Steiner-Khamsi, Stolpe, and Gerelmaa 2004a: 82). Deprived of any meaningful function, most school councils only exist on paper and meet irregularly, if at all.

In an attempt to strengthen community participation in schools, the Rural School Development Project (RSDP), funded by DANIDA, made it a requirement for its project schools to establish school councils, and train their members in democratic school governance. Arguably, the 40 project schools of RSDP host the most active school councils in Mongolia, and yet these school councils encounter a rigid legal framework that leaves little room for meaningful action. In one of the project schools that we visited, four of the five interviewees expressed serious concerns for the lack of transparency at their school. The school council member A5, who as a bank accountant oversees the school budget, maintained a critical tone in the focus group interview (Steiner-Khamsi, Stolpe, and Gerelmaa 2004a: 82f.):

> *A5:* We have no clue [how the money is spent at our school]. But we really should know, I think.
> *A1:* Why? The principal knows.
> *A5:* But I think this is dangerous (*ayuultai shüü dee*), when only he knows and nobody else. There is no control at all (*yamar ch khyanalt baikhgüi*).
> *A4:* I agree, we lack transparency on financial issues.
> *A5:* Exactly, we should have transparency (*il tod baidal*).

Financial contributions by the community had an established tradition, especially for preschools and village (*bag*) schools. During the socialist period, the contribution was made by state enterprises, the animal husbandry collectives, and the state farms situated in the school district. With the dissolution of collectives and state enterprises, the government struggled to find alternative sources of funding. In rural areas with a subsistence economy and a lack of cash, district and provincial governments competed with schools in generating income. In 2000, several school principals reported that they had tried to generate additional income for their school by having their teachers and students, for example, chop and sell wood to the community (Steiner-Khamsi, Enkhtuya, Prime et al. 2000). They planned on using these additional funds, essentially earned through child labor, for purchasing paper supplies and textbooks for their school. Instead, the district and provincial governors, themselves under tremendous pressure to generate additional income and reduce public expenditures, subtracted school earnings from the following year's budget. The more income a school generated, the smaller the budget it was allocated. Needless to say, there was very little incentive for schools to generate income and to raise additional funds from their communities. The practice of subtracting the income raised by schools from the annual school budget is very common, although it was prohibited by the 1995 Education Law (paragraph 33), and confirmed in the 2002 Education Law.

Beginning in 2002, another law (the Public Sector Management and Finance Act), this time designed by the finance ministry, was enacted. The law is diametrically opposed to the general spirit of creating incentives for generating additional income at the school level. In order to centralize control, schools are permitted to have only one account, and generated income must be integral part of the school budget, cementing the fear that schools had all along: that they would no longer have a say in how to spend parental contributions. As a result, how much a school actually collects in donations and fees from parents has become a best kept secret. It is not only unknown to the central levels of government, notably the Ministry of Education and the Ministry of Finance, but it is also unknown to teachers and parents at the school level. It is a secret exclusively shared by the principal, the school accountant, and at times the education manager (assistant principal).

In reviewing the numerous de- and recentralization policies of the past decade, we noticed a host of misunderstandings between the government and international donors, some of them inevitable and others purposefully constructed. For example, several international donors support community participation in schools, and yet the term "community" lacks a semantic equivalent in Mongolia, and is most commonly translated with "home place" or "home area" *(oron nutag)*. School councils are overloaded with officers from the local government (representing the hometown or village) whom Mongolians view as democratically elected community representatives. The international demand for increased "community participation," has another interesting twist which coincidentally sits well with strong Mongolian beliefs in local patriotism. Thus, for example, business people or government officials in Ulaanbaatar who grew up in the countryside see themselves as part of their rural community, and feel obliged to either make donations or channel government funds into their hometown or village.

The confusion continues at each level of administration. The two (world-) systems periodically collide over divergent views on whether schools are *state institutions* or *public institutions*. Mongolian government officials strongly believe that the education system needs to be administered by state representatives, and, as consequence, that principals need to be replaced with each change of government. In contrast, international donors promote the view that schools are public institutions and need to be administered by professionals. A great bulk of external assistance funds was poured into capacity building and the professional development of administrations. It was a poor "investment" as the administrators left leadership positions with the next change of government. Yet, the professionalization of administrators and government officials is placed back on the agenda of each loan, enabling those who signed the loan, and their partisan allies, to make study visits to the West and the East, and personally and professionally benefit from external assistance.

Perhaps the greatest misunderstanding, or more accurately, unspoken disagreement, has to do with definitions of democratic governance. International donors treat democracy and decentralization as Siamese twins, whereas Mongolian government officials strongly believe that an "efficient" and "effective" implementation of any reform, including a decentralization reform, requires (centrally located) "strong hands" and "vigilant eyes." This belief is not entirely unsubstantiated. After all, international financial institutions only deal with government officials, holding them responsible if projects are not implemented or funds abused. Despite a grandiose rhetoric on civic involvement, international donors find it more convenient and efficient to communicate exclusively with government officials. Thus, the lack of social accountability only nominally reflects the hierarchical mannerism of Mongolian government officials. The exclusion of civil society organizations from public policy decisions has to do with how international financial organizations run their business in Mongolia.

Despite the dictatorship of decentralization that structural adjustment policies employ, educational reforms of the past decade wavered on the decentralization issue. If we were to compare the decentralization of educational finance and governance in 2005 to the situation in 1990, we would find that schools received a far greater financial contribution from local (state) sources, and that educational governance was more inclusive of parents than contemporary school councils. Nevertheless, decentralization is held up both by the government and international donors as one of the signposts of the postsocialist period, and functions as conditionality for loans and grants.

Rationalization of Staff and Reorganization of Schools

Mistakes made by international donors carry great weight. The blind push for textbook fees in primary education is one such example. The fee policy was imposed on African countries as well as many others that received loans. Nowadays, it is undisputed that this policy generated a "lost generation" with regard to education. The example of Uganda's fee removal was widely documented in the media (e.g., Dugger 2004) because it instantaneously led to huge surges in enrollment. The fee policy was eventually lifted after severe criticism from within the World Bank (e.g., Bentaouet Kattan and Burnett 2004). Heyneman (2003) provides several other examples of ill-conceived, and yet forcefully imposed World Bank policies, most of which were imbedded in a narrow and neoliberal conception of human capital theory.

In Mongolia, there is a long list of misguided policy decisions that, after a period of experimentation, were reversed. Examples include the following:

- Script reform (1992–1994)
  The reform attempted to replace Cyrillic with the Mongolian script, which had not been in use for over 50 years (Cyrillic was introduced in 1941). Because the Ministry of Education did not succeed in attracting sufficient international funding, the teachers were unprepared, the textbooks unavailable, and parents up and arms against this reform. The reform was suspended after only two years.

- Dormitory fees (1996–1999)
  Students were only accommodated in school dormitories if their parents provided meals. This policy, also known as the "meat requirement," had a tremendous impact on dropouts and non-enrollment (see chapter 9). The policy was dismantled three years later.

- Rationalization of staff and reorganization of schools (1997)
  Schools were "rationalized," that is, the number of school staff was drastically reduced, and schools were either completely shut down or reorganized in 1997. The reorganization pursued three strategies: (1) to close down small schools in remote rural villages (*bag* schools), (2) to discontinue grades 9 and 10 in rural schools (*sum* schools), and (3) to merge schools in cities and province-centers into large complex schools [*tsogtsolbor surguul'*]. What are the remnants of the 1997 policy a decade later? The schools in remote villages (*bag*) are in ruins: the few that survived became either administratively affiliated with a school in the same rural district ("dependent *bag* school") or struggle to maintain their status as a community-based, independent *bag* school. The *sum* schools subverted the closure of their ninth and tenth grade, and eventually, with support from provincial education authorities, reinstated the two terminal grades. The complex schools finally, continue to exist, but are criticized as gigantic educational fortresses that are not only difficult to manage but also pedagogically non-conducive for younger students.

In this chapter we only comment on the rationalization of staff and the reorganization of schools. These measures were approved in 1996, and were implemented in 1997. We cannot overemphasize the importance of semantic nuances throughout this book: Rather than claiming—as it is frequently done—that these measures were funded with the support of a large loan ($15.5 million) from the Asian Development Bank, it is more accurate to state that the Ministry of Education was given large loans *for* rationalizing staff and reorganizing schools. The reform was part and parcel of the first Education Sector Development Programme (ESDP) which, according to the Ministry of Education, targeted the following:

- Rationalize education structures and staffing
- Promote cost recovery schemes
- Support privatization and private provision of education, and
- Develop a comprehensive policy framework for technical education and vocational training (MOSTEC 2000: 2)

The rationalization program was orchestrated as a campaign. To set the tone, the Ministry of Education fired more than 8,000 school staff in 1997. Supposedly, most of them were support staff, not teachers. Few believed in the "rationalization of teaching and non-teaching staff ratio" (MOECS and ADB 2001: 59), and the media reported that many experienced teachers were either dismissed, or voluntarily stepped down in protest. Moreover, among the support staff were key personnel such as dormitory staff who cooked, cleaned, and looked after the dormitory students. School librarians and school herders in charge of the school's livestock were also categorized as secondary. Misguided by a Western image of a school, these "support" services were seen as nonessential to the functioning of a school. The severe cuts made the public painfully aware that the language of the new allies was one of structural adjustment. The Democratic Union-led Ministry of Education (1996–2000) rigorously enforced the formula—reducing expenses and generating income—in all aspects of the education sector. Not surprisingly, the Ministry of Education was chastised for bending under international pressure and enacting such harsh measures.

The reform was pronounced in a climate of widespread dissatisfaction among teachers. Two years earlier, 6,000 teachers went on strike for 143 days (Chimeddorj 1997: 2) demanding higher salaries. The strike, or—literally translated from Mongolian (*ajil khayakh temtsel*)—the "throwing-away-work" rally started out with a 1,000 senior teachers putting down their work; within a few days, the operation of schools came to a complete standstill. The Teachers Union appealed to the United Nations High Commissioner for Human Rights, and requested international support "to speedily stop the rapid impoverishment of teachers" in Mongolia, to triple teacher salaries, and to adjust salaries to the inflation rate (Confederation of Mongolian Trade Unions 1995: 1). The 1995 strike was backed by huge crowds that rallied on the main square in Ulaanbaatar, blocking the entrances to the parliament and government buildings. Two years later, anticipating the eruption of new public protests, the Ministry of Education rapidly carried out the rationalization reform shortly before summer break. The Teachers Union prompted the decision with a one-day warn strike in June 1997. Seven thousand teachers, again supported by thousands of civilians, participated (Chimeddorj 1997: 2). The Teachers Union accused the Ministry of Education of having prepared the ESDP behind closed doors. During a press conference the Minister of Education wryly commented, "Teachers don't trust us!" (Stolpe 2001: 72). According to the Minister of Education, the culprits were not the teachers but the principals. As government representatives, the principals were responsible for convincing teachers, and obviously had failed to convey the necessity of a shock therapy of sorts for schools.

The storm of protests achieved little. The Ministry of Education continued on the route of structural adjustment. The fact that it "discharged surplus staff" was notched up a success (MOECS and ADB 2001: 60), and summarized in a language more commonly used by financiers than by educational experts: "Thus, this amount [T 3.3 billion] is saved in the education sector, or in other words, investment made in the educational sector equals to this amount."

The next step was the reorganization of schools. Similar to the rationalization of staff, the reorganization of schools was a comprehensive reform targeting the village level (*bag*), rural district level (*sum*), as well as province-centers and cities.

Closing of Small Village Schools

By far the most blind-sighted reform was the reorganization of schools in rural areas, notably the abolishment of independent village schools (*bag* schools). During socialist times they were in large part funded by the animal husbandry collectives (*negdels*) and had strong community participation. Their location in remote rural areas enabled children from herder families to access a school in their vicinity.

In the early 1990s, families deserted remote rural areas because the collectives were shut down, and families were not able to make a living from the heads of livestock that they received in the wake of privatization; either because their herd rapidly diminished, or they never got as many heads as they were entitled to. The situation at the end of the decade was quite different. Those who managed to prosper or at least get by with animal husbandry remained in remote rural areas. Both the school and the health post, visited by a nurse every couple of weeks, were vital for these families. Starting out with 1 teacher per grade at the beginning of the decade, the *bag* schools were first transformed into multigrade primary schools with 1–3 teachers, and then completely phased out. The abolishment of small village schools, along with the introduction of the dormitory fee ("meat requirement"), had a devastating impact on rural development. Schools in rural Mongolia do not only serve children but also operate as cultural centers for the community. In fact, they are the heart of community life, and the center of rural infrastructure. The dissolution of the small village schools became a push factor for migration, and it was the main cause for rural flight in the late 1990s.

At the beginning of the 1990s there were over 1,000 *bags*, and hundreds of them had their own school (Government of Mongolia 1993: section 1–15). In school year 2004–5, there were exactly 53 *bag* schools left that offered grades 1–4, and operated as independent village schools. In addition, there are a few small village schools that are financially linked with schools in their rural district. Nobody knows their exact number because these "dependent *bag* schools" have merged administratively with rural district schools, and they are not separately listed.

Termination of Grades 9 and 10 in Rural District Schools

Over the past 15 years, the structure of the educational system has transformed in 2 ways: primary school has been extended from 3 to 5 years, and the school curriculum has been extended from 10 to 11 years (see also chapter 2). The extension of primary school from three to five years was undertaken incrementally. The previous 3 + 5 + 2 structure—3 years primary, 5 years lower secondary, 2 years upper secondary—was replaced in 1996 with a 4 + 4 + 2 structure. Then in 2004—in line with the curriculum extension from 10 to 11 years—the school entrance age was lowered from eight to seven. Since 2004, primary school lasts 5 years, lower secondary 4 years, and upper secondary 2 years (5 + 4 + 2). As a result of the curriculum extension, compulsory education was prolonged from eight to nine years. For structural adjustment reformers, the pressing question became where, within the given structure, savings could be made. The options were laid out in the first Education Sector Review:

> Generation of additional resources should be accompanied by efforts to reduce costs. One way to do this is to close either parts of institutions or whole institutions. There have been discussions of closing grades 9 and 10 in some schools. (Government of Mongolia 1993: section 1–33)

As mentioned before, the majority of *bag* schools were shut down and, beginning in 1999, the *sum* schools (rural district schools) became a target for "structural rationalization." The majority of *sum* schools were forced to close their ninth and tenth grades. At the same time, a few *sum* schools were elevated to regional schools. These so-called *inter-sum* schools were supposed to accommodate eighth grade graduates from other schools and serve as a resource and in-service teaching center for surrounding schools. They were intended to act as regional schools, but did not for a variety of reasons. First, they neither had the staff, facilities, nor budget to expand their own ninth and tenth grade classes to accommodate newcomers from the surrounding schools. Second, the *inter-sum* schools began to see themselves as less rural and far superior to all their surrounding schools. Many of them established separate tracks or separate classes for students coming from the "countryside" for fear that their quality standards would be lowered by these intruders; a few completely refused to accept any new students (Steiner-Khamsi, Enkhtuya, Prime et al. 2000).

A war broke loose among parents. Under tremendous pressure from parents at their school, school directors gave preferential treatment to eighth grade graduates from their own school, and quickly filled the few available seats in ninth and tenth grade with their "own" students. Faced with a dead-end situation, parents from regular district schools soon realized that the only way to secure a place in a ninth or tenth grade class was to enroll their child in an *inter-sum* school before eighth grade; they hoped that their child would be considered a student of that school and therefore be admitted to upper secondary school. The *inter-sum* schools started to fill up rapidly after fifth grade, and the situation became intolerable for all parties involved. A counterreform movement from "below" emerged. Parents, teachers, and school administrations filed countless requests with the Education and Culture Departments to reopen the ninth and tenth grades in their school. In an act of subtle subversion, the ECD directors approved these requests individually. In July 2004, a ECD director from a Western province in Mongolia told us with a sense of accomplishment (Steiner-Khamsi, Stolpe, and Gerelmaa 2004b):

> We approved all the re-transformation requests, and now we don't have anymore grade 1–8 schools in our province. All our schools are again grade 1–10 schools. What were they [central government] thinking, when they closed down grades 9 and 10? The parents were running down our office to undo this unjust reform.

The last MPRP-led Ministry of Education (2000–2004) was generally supportive of rural school development, and tolerated the reversal of a reform that was approved under the previous administration. In school year 2004–2005, there were 681 schools in Mongolia (MOECS 2005), 98 (15 percent) of which are private schools, 3 are schools for children with disabilities, and a few are specialized schools (music schools, university preparation schools, etc.). The public schools (excluding private schools, schools for children with disabilities, and specialized schools) fall into the following three categories:

1. 53 primary schools (grades 1–4)
2. 167 lower secondary schools (grades 1–8)
3. 351 upper secondary schools (grades 1–10)

Grade 1–8 schools have been gradually reconverted into grade 1–10 schools. In 2004–2005 only every third secondary school was a primary (grades 1–4) and lower secondary school (grades 1–8), and it is expected that the pressure to reconvert the remaining 167 schools will persist.

Complex Schools in Province-Centers and Cities

The newly established complex schools offer all grade levels and host on the average 3,000 students. This is how these giants were created: several schools in an area were merged into one large school. In most locations the merged schools were within walking distance, but in some places (e.g., in the Arkhangai province-center) they were located at a great distance from each other. The idea was to save money by firing "surplus" administrative staff, and by introducing more cost-effective school infrastructures in which the "surplus" laboratories, libraries, concert halls, physical education halls, and so on were eliminated, and concentrated in only one of the school buildings. In an educational context where there is a shortage of almost everything including school facilities, it is cynical to draw on a concept of "surplus." Unwaveringly, the Ministry of Education went ahead and established 40 complex schools. These mega schools are known as the "ADB schools" because they received generous funds from ADB to equip their laboratories, computer rooms, and libraries (see MOECS and ADB 2001: 60ff.). They were able to do so at the expense of all surrounding schools. The establishment of well-equipped complex schools in the province-centers generated a two-tiered education system: 40 rich (ADB) schools and more than 500 regular, poor schools.

Ironically, during the same period in which large schools fell out of fashion in high-income countries, the Government of Mongolia converted well-functioning small schools (including primary schools) into large complex schools. In the United States, for example, the big-is-beautiful belief of the 1970s has faded away, paving the path for a forceful small school movement that is rapidly gaining a foothold in city schools. In Mongolia, only seven years after this sweeping reorganization, the blind enthusiasm for complex schools also experienced a setback. The glimmer of hope is that the evaluation of complex schools, as requested in the third Education Sector Program (funded by ADB), will correct the mistakes of the reorganization reform.

To reiterate the ECD director's rhetorical question: What were they thinking? Clearly, even with a far stretch of imagination it is difficult to comprehend why a ministry of education signs off on a loan that curtails educational attainment. The global trend moves in the opposite direction: more years of schooling for more individuals. The structural adjustment reforms of the 1990s have placed Mongolia in a peculiar web of inverse social mobility. Typically, each generation is more educated than the one that precedes it, a phenomenon commonly regarded as one of the indicators of modernization. Viewed from such a broad sociological perspective, the modernization process in Mongolia was essentially reversed in the 1990s: the current generation of adolescents is less educated than their parents' generation because they lack access to schools; either they never enrolled, dropped out, or only completed eight years of schooling.

In contrast to many other reforms that were voluntarily borrowed, the structural adjustment reforms discussed in this chapter were clearly imposed by the Asian Development Bank. Strikingly, as nonnegotiable these reforms appeared to be, a few of them were subverted or dismantled. Most reforms of the 1990s, however, were irreversible and greatly have contributed to widening the social gap between rural and urban areas, and between the poor and nonpoor population.

6

THE MONGOLIZATION OF
STUDENT-CENTERED LEARNING

In June 2001, we visited a school in the province of Khovd that for the past three years had been experimenting with student-centered learning (Steiner-Khamsi 2001). This particular visit is memorable because it inspired us to want to understand better the structure and culture of the Mongolian classroom.

The school was one of 72 partner schools in the School 2001 project. As in the previous two annual project evaluations, we visited partner and non-partner schools across the country, submitted reports to the Mongolian Foundation for Open Society, and discussed suggestions on how to improve the design of the project for the following year. The evaluations were based on surveys, interviews, and observations. With each visit we were determined to follow the Mongolian script for evaluation visits: Usually a one- to two-hour meeting with the principal and the education manager (assistant principal) in which the visitors listen to a long list of facts about the school,[1] and then are given the opportunity to ask questions for clarification. This is wrapped up with a walk through the school facilities. The second part takes place in the teachers' room where instructors display their teaching portfolios, share their experience with the "new teaching technologies" (*shine surgaltyn arga*)—the Mongolian term for student-centered learning—and request feedback on their lesson plans, teaching materials, and booklets. Finally there are classroom observations, followed by individual meetings (interviews) with teachers. Since every teacher in the school wants a chance to meet with visitors from outside of the province, several attempts are made—and regularly rebuffed—to pull us out of the classroom after only 10 or 15 minutes. Not all foreigners submit to the etiquette of school visits, even though the rules are straightforward and reasonable: visitors are supposed to first listen, look, and learn, and only then ask questions and make comments.

Nothing was out of the ordinary during the first part of our visit. After a long tour of the school facilities, the teachers welcomed us in the teachers' room. Unlike on previous occasions, the teachers did not inundate us with a show and tell of lesson plans and teaching materials, but instead they signaled their interest in moving directly to the meeting portion of the visit. Something very special had happened the past week, and they wanted to tell us about it. Four teachers—the Mongolian language teacher, the geography teacher, the history teacher, and the social sciences teacher—had worked together to prepare a lesson plan on a topic from the history of the Mongolian Empire. The lesson was cross-disciplinary, covering material from all four subjects. They co-taught the topic over a period of five sessions, and merged the four classes into two classes of students. They began the last session by making students gather their questions in small working groups. Afterwards, representatives of the working groups read the questions in class, and everyone—teachers and students—responded to the questions. The story ended with an enthusiastic account of the overwhelmingly positive reactions from both the teachers and the students who had participated in their experiment.

The two Mongolian colleagues in our evaluation team immediately identified it as a major breakthrough in student-centered learning. Why were we so unimpressed? After reviewing the details of their account, we realized that we had focused on the wrong parts of the story. What caught the attention of the Mongolian colleagues was not so much the "integrated curriculum" (encouraged in School 2001), making teachers work across disciplinary boundaries, or the fact that they had lectured to over 60 students (discouraged in School 2001), but rather that they had started the last session "cold." That is, the students had been encouraged to gather questions for which the teachers risked not having a ready answer. Furthermore, the boldness of their pedagogical approach was also apparent in the invitation to students to take part in answering questions. It took us another two years to understand that the students who read the questions for their group in class were not regular students, but either row, group, or table monitors.

Our perplexity that day made us realize we were worlds apart from their notion of student-centered learning. In the years that followed, we kept an eye out during class observations for any other occasion that might present itself as a point of entry for understanding the deep social structure and fine cultural texture of the Mongolian classroom. Two small empirical studies—one on pedagogical jokes, and the other on class monitors—helped us to partially lift the mystery of the Mongolian classroom in a more systematic manner.

What Teachers Find Funny

Our study on pedagogical jokes was the product of another project we had been forced to abandon because the research question was irrelevant. The project was meant to explore teachers' classroom management and disciplinary techniques. What we didn't realize, however, was that in Mongolia management and disciplinary issues are delegated to students in the classroom. The class monitor effectively functions as an assistant teacher and takes responsibility for student conduct. So when we asked the teachers about their own management style they didn't understand what we were referring to. We followed up on this discovery in a separate study on class monitors, which we present later in this chapter. But since the original research question generated a lot of confusion and made our interviewees feel uncomfortable, the teachers distracted themselves by telling jokes. On our second day of data collection in the bilingual province of Bayan-Ölgii (Kazakh and Mongolian), two members of the research team took notes on what the interviewees had to say on classroom management and disciplinary techniques, and the third member focused exclusively on jotting down their anecdotes (Steiner-Khamsi, Myagmar, and Sum"yaasüren 2004). Our transcripts include the funny stories that school teachers (N = 10), teacher education students (N = 35), and lecturers of teacher education (N = 26) either shared with us or among themselves.

There is a vast body of academic literature on jokes ranging from classic psychoanalytical theory (Freud 1965) to ethnographic studies on black humor and its function in the practice of storytelling (e.g., Goldstein 2003). What often makes jokes funny is that something very private and personal is made public.[2] Our investigation of pedagogical jokes is an attempt to bring to light the "dark" professional secrets that educators are supposed to keep to themselves, but of course eagerly share whenever an opportunity arises, generating laughter and affirmation. The stories are usually embarrassing, or filled with anxiety or shame. Cracking jokes with other educators is a form of professional communication; and a very effective one at that because it releases thoughts and feelings that a group of people has in common. Precisely because jokes reflect and reproduce shared experiences, they are not necessarily humorous to individuals outside the community. Thus, episodes that are funny to Mongolian teachers are probably not interesting to Mongolians in other professions, and their sense of humor is quite different from what people in other countries would consider funny.

Our inventory of pedagogical jokes revealed a central theme: The ignorance of the teacher. Mongolian educators find stories about teachers who don't know the answer to a student's question uproariously funny. We confine our description to two examples:

Third-Year Student at the Teachers College, Bayan-Ölgii:

During the teaching practicum, one student asked me in front of everyone: "The male goat eats standing up, and female goat eats lying down. Why is that?" I had no idea what to say, because I had never asked myself such a question. Then, I said: "I am sorry, but I don't know the answer to this question." You should have seen the expression on the student's face: he was totally shocked that I didn't know the answer.

Practicum Coordinator, Bayan-Ölgii:

One of the student teachers was asked a question and she didn't know the answer, but her solution to the problem was creative. She said to the class: "This is the assignment for tomorrow's lecture. Please ask your parents, and come back with the correct answer tomorrow!"

The audience roared with laughter in response to these stories because they undermine and oppose what a teacher in Mongolia is first and foremost supposed to be: knowledgeable. The narrative of the pedagogical joke starts out with the speaker admitting ignorance, but then goes on to describe how she/he managed to avoid looking foolish. Arguably, what is considered a pedagogical faux pas in Mongolia does not necessarily qualify as embarrassing in other cultural contexts. Though we haven't systematically studied pedagogical jokes in other countries, we have noticed that teachers in Central and Western Europe, and in the United States, tend to joke about their inability to control the class. The central theme of these stories tends to deal more with student discipline, and less with the teacher's inability to answer a question. In Mongolia, the emphasis on knowledge as the key qualification of a teacher permeates all aspects of the profession.

Throughout the 1990s, the state-run in-service training for teachers focused exclusively on the content of what was being taught, with little regard for pedagogical issues. However, starting in the mid-1990s international organizations established a parallel in-service training system that offered courses on critical thinking, developmentally appropriate teaching methods, debate, student-centered learning, cooperative learning, and interactive teaching methods. The Mongolian Foundation for Open Society, Save the Children UK, and DANIDA funded such courses, and they were notably popular with teachers. Most of these international projects were Train-the-Trainer projects, that is, international workshop moderators trained a select group of Mongolian teachers, who were eventually certified as in-service teacher trainers (see also chapter 8). This so-called Cascade Reform Strategy was most systematically pursued in the School 2001 project where three to five teachers per project school ("core team") were trained as peer trainers and mentors, and acted as workshop moderators for school-based in-service training. As with other imported practices, Mongolian educators selectively borrowed elements of the new ideas and practices they were exposed to. Since reform of curriculum standards lagged behind the pedagogical innovations in schools, teachers found themselves in an awkward bind. As a teacher from a School 2001 partner school commented:

> [A]t our school we frequently use interactive teaching methods from September through March. In April, we refocus on content so that our students pass the exams in June. (Steiner-Khamsi, Enkhtuya, Prime et al. 2000: 57)

The great bulk of teacher performance measures tend to emphasize content over method. The teacher olympiads or competitions, a popular relic from socialist times, exclusively assess teachers' knowledge on a given subject matter. Accordingly, teachers are awarded at English language olympiads for knowing English, not for teaching English effectively.

Nevertheless, several attempts were made at the institutional level to broaden the key qualifications of teachers. For example, the Mongolian State University of Education, alma mater of over 70 percent of all teachers and school administrators in Mongolia, launched a comprehensive curriculum reform in 2002. The reform aimed at an integration of pedagogical knowledge and skills, a better organized teaching practicum (the placement of teacher education students in schools), and an inclusion of practitioners and practicum mentors as "clinic professors." At the school level, most teachers have been exposed to or at least heard about student-centered learning techniques. After a decade of experimentation with Western style teaching methods, there is a proliferation of Mongolian variants, or hybrids of student-centered learning, that all have one feature in common: the imported teaching methods are recontextualized in ways that suit the hierarchical order in the Mongolian classroom.

Female, Smart, and a Leader—The Class Monitor

In the Mongolian classroom teachers mostly lecture while the students listen and take notes. Every now and then the teacher turns to the class and asks a question, and sometimes walks through the three rows of students and comments: "This is correct, my child!" or "Redo this page, my child!" The "child" never addresses the grown-up directly, but rather waits until it is given the permission to talk. Learning takes place in an atmosphere of utmost respect toward the teacher, and teachers in turn are expected to be respectful and kind toward their students. The bond between teachers and students lasts a lifetime, and often Mongolians identify themselves as the followers or students (*shav'*) of former school or university teachers (*bagsh*).

A wide gap exists between the status of the teacher and the status of the student, similar to the relationship between adults and children in general outside of the school. However, the class monitor (*angiin darga*) bridges this gap. Elected by the students, the monitor acts as assistant teacher, organizer of events, and discipliner of peers. The Mongolian classroom would be a different place without the class monitor, and the class monitor system is one of the distinct features of Mongolian pedagogy.

The practice of selecting a few students as role models, leaders, and mentors has been a long-standing tradition in Mongolia. It is on the one hand a legacy of the socialist past, reflecting the principle of "democratic centralism" whereby a few, elected members of society, assume leadership roles within a collective. This pyramidal structure was present in all aspects of socialist life, both in Mongolia and in other socialist countries, and it was replicated in the micro-collective of the Mongolian classroom. A popular topic at international conferences of socialist educators was the principle of student self-governance, and the specific role of class monitors (e.g., Changai 1974). The concept of student self-governance drew on Makarenko's influential work on children colonies. In the context of schools, in particular, these class monitors had to be outstanding in every regard—academically, and also in terms of communication and organizational skills.

However, it would be wrong to highlight only the socialist tradition when explaining the hierarchy in the Mongolian classroom. Similar to many Asian countries, Mongolian society is structured more by age than by any other social category. Prior to the introduction of the class monitor system, the oldest child in the classroom assumed the role of class leader. In contrast to the class monitor system, the oldest in the class received their position by default, rather than by a democratic process. Throughout the socialist period both systems existed side by side in the Mongolian classroom, merging into a hybrid model: the socialist system of an elected class monitor, along with the cultural tradition of class oldest. Only in the last few years have the oldest in class lost their status as sources of authority, and in fact are sometimes stigmatized.[3] For the most part however, teachers, school administrators, and even the other students still treat the oldest child in the classroom with special respect. Continued high esteem for the class oldest can be explained as an extension of what children experience at home and outside the school. There is a gender-neutral term for the oldest child in the family (*uugan*). *Uugan*s not only have special privileges and obligations toward other family members, but they are also expected to act as a role model for younger siblings. Belief in seniority as a principle structuring interaction is reflected in the proverb (Stolpe 2001: 6). Every human needs an older person on top as much as the *deel* (traditional Mongolian clothing) needs a collar (*khun akhtai, deel zakhtai*).

Our study on class monitors was conducted in April and May 2004, and included individual and group interviews, questionnaires, and diary entries of former class monitors (Steiner-Khamsi and Kuliyash 2004; see also chapter 1). The Mongolian coresearcher was O. Kuliyash, lecturer of education at the teachers college in Bayan-Ölgii. Kuliyash kept her own class monitor diaries from when she was a child, and for the past 20 years collected and analyzed the notes passed on to her by class monitors she has taught. The survey was administered to 86 former and current class monitors (youngest: 9, oldest: 52). Grown-ups serving as class monitors before 1990 were purposefully included in the sample to provide a historical perspective.[4] Even though the sample was not randomly selected,[5] nor did we attempt to reflect the experiences of Mongolian class monitors nationwide, our study draws on much more "data" than is recorded in the research project. Once word spread that we were interested in understanding the class monitor system, Mongolian colleagues and friends spontaneously shared their own childhood experiences as class monitors. These numerous informal conversations have shaped our understanding of the system as much as the data that we collected from the study.

Not surprisingly, over 80 percent of the respondents in our study were female. This overrepresentation corresponds to the fact that in Mongolia, the best students in the class are typically female. One-third of the former class monitors were class monitors for the entire duration of their school careers, from preschool to higher education. The class monitor has several assistants, and on average spends three hours a day after school supervising assistants, disciplining students, coordinating the cleaning and repair of the classroom, organizing events, and conducting individual and group tutoring.

The pyramidal hierarchy of the Mongolian classroom is represented symbolically in the Figure that follows. The class monitor (assistant teacher) is

Figure 6.1 The Hierarchical Setting in the Mongolian Classroom

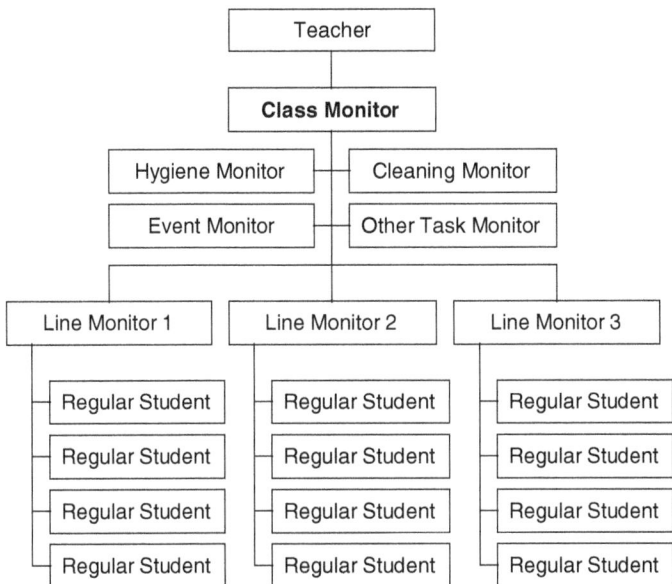

"on top" in the front row, and the regular students, traditionally sitting in three rows at the back, are on the bottom. In the middle of the pyramid are two types of assistants: task monitors (hygiene monitor, event monitor, cleaning monitor, etc.) and line monitors. In traditional classrooms the line monitors consist of three row monitors, while in classrooms focused on student-centered learning there are four to six table monitors. Figure 6.1 illustrates the hierarchical setting in the Mongolian classroom.

A few comments on task and line monitors, and their relationship to the class monitor, might be useful for understanding Figure 6.1.

Line Monitors

In most classrooms, the student desks are set up in three rows, each of them managed by a row monitor (*salaany darga*). In contrast, teachers that have been exposed to interactive teaching methods tend to replace rows with group tables (two student desks per group accommodating approximately six students), ending up with four to six groups and four to six group monitors (*bagiin darga*) or table monitors (*shireenii darga*). The line monitors are exclusively responsible for the academic and disciplinary issues of the students in their row or group, and report to the class monitor.

Task Monitors

The class monitor appoints students to be in charge of specific tasks, depending upon their grade level. We found in our study that hygiene monitors, responsible for checking clothes and inspecting the cleanliness of fingernails, hair, and face, are found up through higher education. The task of organizing after-school events (picnics, sport events, concerts, competitions, parties, etc.) grows with each grade level, and often class monitors in upper secondary school and higher education appoint an event monitor to assist them in what becomes a monumental task. According to our survey, the two most dreaded responsibilities of class monitors are the coordination of classroom cleaning and collecting money (fees or donations) from parents. The class monitors try to delegate these two tasks to other students, and are apparently not always successful in finding volunteers willing to take them on.

The following Figure 6.2 presents the main tasks of the class monitors, and shows the percentage of overall monitoring time (average three hours per week) allocated to each task. The Figure 6.2 distinguishes between the class monitors in the sample (N = 86) who serve at the primary school (grades 1–4), secondary school (grades 5–10), and higher education levels.

Disciplining students in and outside of the classroom is a major task for all class monitors, and especially for those serving at the primary and secondary school levels. It absorbs approximately 20 percent of the time that a student spends monitoring the class. As the following three excerpts from our interviews illustrate, disciplining students has a different meaning at each school

Figure 6.2 Percentage of Overall Monitoring Time Spent on Various Tasks, by School Level

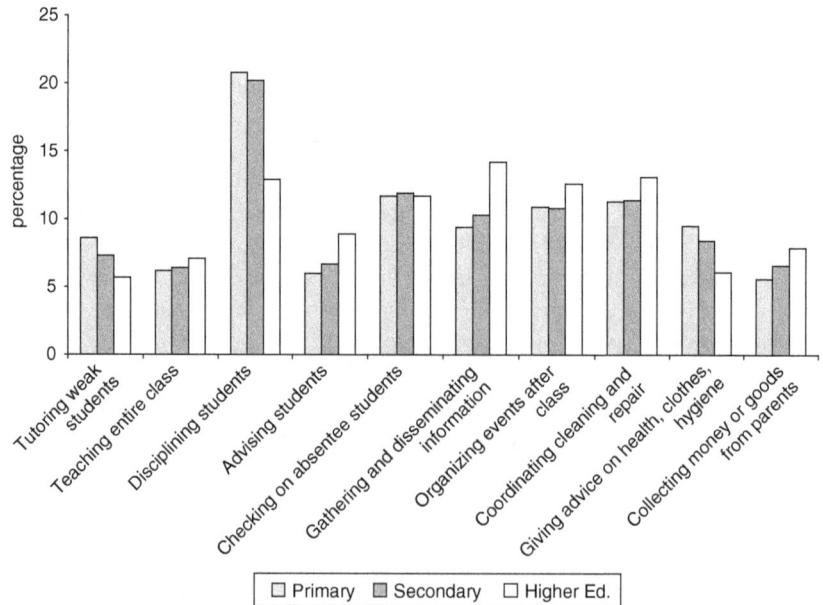

level (Steiner-Khamsi and Kuliyash 2004: 78):

> Grade 1 class monitor, 9-year-old female:
>
> I find the most rewarding activity as a class monitor is to discipline students who sit in the wrong position. I feel very good when they learn from me how to sit correctly.
>
> Grade 7 class monitor, 14-year-old female:
>
> As class monitors, we have to work with every student in the class, and make sure that none of the students falls behind, or feels excluded. Then, we have to discipline students who have a bad attitude towards learning, the teacher, the students, or the school.
>
> Year 2 college student, 20-year-old male:
>
> At first, I had problems getting respect from difficult students. Students who are impolite towards the teacher usually also speak back to the class monitor. But then I spoke several times with the disruptive student in private, and explained to him that his behavior is not only bad for his own professional career, but also had a negative impact on the entire class. Gradually, he started listening to me, and accepted my advice.

Diaries of class monitors read like the notebooks of psychotherapists. They contain observations of students in class, and for those students that appear to be troubled the class monitor notes her diagnosis of the problem, her intervention strategy (advice), and the response of the student. Talking their peers into participating more actively in classroom activities, and convincing them to take responsibility for others in class, are often seen as solutions to the problem. Their notes tell a story of patience and compassion, but also of pain when all the support and advice they've given have still failed to reintegrate the troubled student. Class monitors of the pre-1990 era reported how they would visit families when a peer was absent for a few consecutive days, making sure that the student did not drop out of school. Besides being a confidante for students and a liaison between teachers and students, class monitors are expected to assume leadership roles. The most frequent responses to our open-ended question as to why class monitors are needed, were "everyone needs a leader," "peers understand each other better," and "someone needs to make decisions on behalf of a group"; without a leader, "there would be chaos," "the class would fall apart," and "nobody would work for the common good of the class" (Steiner-Khamsi and Kuliyash 2004: 76). Our study on class monitors enabled us to explore the internal dynamics of the classroom, and to investigate the social order within this microcosm.

STUDENT-CENTERED LEARNING IN A HIERARCHICAL CLASSROOM

One of our unexpected discoveries was that teachers only introduced interactive teaching methods at a particular point in the lesson (Steiner-Khamsi,

Enkhtuya, Prime et al. 2000: 55ff.). The teachers we observed in the School 2001 partner schools tended to organize their lessons along the following sequence: (1) repetition of the previous lesson, (2) introduction of the new lesson, (3) responding to student questions for clarification, (4) independent student work on the new lesson, and (5) announcement of homework.

Apart from a few experienced teachers who completely revamped their lessons after recognizing the experiences and knowledge of students and their families as a valuable resource, most teachers in the School 2001 project chose to only modify the third sequence of their lesson. Instead of having students ask questions individually, they made them gather all questions in small groups first, and then present them, group by group, to the class. Several teachers invited the other groups to join the question and answer session, and actively encouraged them to come up with answers. As unspectacular as this may seem to outside observers, the opportunity to discuss a topic with peers was quite remarkable for the students. Allowing students to speak at a time they usually didn't was regarded as a major breakthrough in student-centered learning. An important byproduct for the teachers was that they had "time off" from lecturing during students' group work, and many of them used that time to communicate—sometimes for the first time—with individual students in class.

In the first few years of the School 2001 project our attention was directed exclusively at the teachers, and on how they locally adapted, or "Mongolized," imported constructs of student-centered learning. But what about the assistant teachers—the smart and for the most part female—class monitors? How did their role change as a result of an externally induced change in pedagogy, and how did the Mongolian variant of student-centered learning affect interactions among students? After being sensitized to the hierarchical order of the Mongolian classroom, we stopped focusing on how the teachers interpreted student-centered learning, and instead turned our attention to the dynamic in classrooms that were exposed to "new teaching technologies."

As we elaborate in greater detail in chapter 7, policy makers and practitioners and educators are fond of the "scientific approach" to education that, in practice, translates into formulas, indicators, and benchmarks to measure the quality of education. A teacher education student in Mongolia, for example, learns to observe whether at least 60 percent of the students—understood as the normative threshold for effective teaching—participate in the lesson. This particular measure, established by lecturers of teacher education, overshadows all other indicators of student participation; teacher education students must apply it religiously. It became important to the Mongolian colleagues on the research team to come up with percentages, and the burning question was whether student-centered learning increases or decreases student participation. After several discussions, the research team agreed to also include a focus on student-student interaction.

B. Sum"yaasüren was an experienced educational manager (assistant principal) at the Shav' Complex School in Ulaanbaatar and adjunct professor at the Mongolian State University of Education ("clinic professor"). She developed an instrument for getting at percentages by counting the number of students who responded to the teacher's questions, and divided that number by the total number of students in the class. Kh. Myagmar, a renowned psychology lecturer at the Mongolian State University of Education, and G. Steiner-Khamsi focused their classroom observations on students who broke the rules: Students who were given the floor but didn't respond, students who interrupted other students who had been given the floor, and students who were obviously not paying any attention to the lesson (sleeping, talking with other students, or engaging in an activity unrelated to the lesson). As a team, we observed several classrooms in Bayan-Ölgii and Ulaanbaatar, and compared our notes after each visit. In addition, Kh. Myagmar and B. Sum"yaasüren had their teacher education students use the instruments and rosters to collect additional information for our research questions.

Three of the most important findings were (see Steiner-Khamsi, Myagmar, and Sum"yaasüren 2004): First, a single percentage figure didn't capture the full range of student participation, as some students (class and line monitors) spoke up to 10 times in a 50-minute lesson, whereas others spoke only once. The students who were most active in class were, without exception, the monitors. Second, the teachers always used assistants for the group work. The job of these assistants was to moderate the group discussion, and then report on the group work to the class. The students who were chosen as moderators were, again, the group or table monitors. Third, disruptive students were dealt with by those sitting near them and the monitors, rather than the teacher. These findings are interrelated in that monitors play an active role as assistant teachers and in disciplining their peers.

In other words, group work—one of the new practices introduced as part of student-centered learning—did not enhance direct communication between teachers and students in the classroom, but it did increase student-student interaction. Teachers communicated with students via the group monitors and vice versa: students remained at the same place in the hierarchy, but now relied upon group monitors to respond on their behalf. This is not to suggest that regular students were not given a chance to speak. In fact, in many lessons, the monitor or teacher asked the other students in the group whether they had something to add to the wrap-up presented by the group monitor. Only a few took the opportunity, and the majority of students were content with being represented by the monitor. In classes that were committed to student-centered learning, the presentations, discussions, and other activities remained teacher-led, and group work, in turn, was still led by the monitors. At no time during the lesson was there room for student- or group-initiatives, or student-led activities.

While we found it fascinating to observe how group work plays out in the hierarchical setting of a Mongolian classroom, not all Mongolian parents or teachers are enamored with the system of class monitors, and some of them find the system—for ideological or psychological reasons—to be outdated. Ideologically, the class monitor system is blamed for the so-called *darga*-mentality manifested in passivity, lack of critical thinking, and blind obedience. By the time Mongolians have completed school or higher education, they are used to relying at all times on a leader (*darga*) who tells them what to do, and what not to do. Moreover, the issue is not only that regular students revere class monitors, but also that class monitors report on their peers to the teacher. The chain of command continues at each level, making the individual vulnerable to denunciations from peers.[6] Although many ambitious parents attempt to win the teacher's support for having child elected as class monitor, as a first step toward becoming a leader later in life, we witnessed lively conversations in which the psychological benefits were questioned. There is a wide spectrum of opinion, from those who find the system cruel for the students who are never elected, and consequentially suffer low self-esteem, to those who fear that leadership fosters inflated egos, which will later be shattered.

Although this chapter only deals with the hierarchical setting in the classroom, we would like to present an illustration of the hierarchical order within schools (see also chapter 7). In one of the rural district-schools in Övörkhangai, we came across a portrait of Lenin in the hallway of the school. Lenin—in socialist student textbooks referred to as Lenin *bagsh* (Lenin Teacher)—is surrounded by teachers and two students, all dressed as (communist) pioneers (photograph 6.1). The painting was placed in the school entrance hall in the 1980s, and teachers were reluctant to remove it because the majority of teachers depicted in the painting are alive, and still work as

Photograph 6.1   Lenin *Bagsh* with Teachers from a Rural District-School, Övörkhangai
Source: Photograph by Ines Stolpe and Gita Steiner-Khamsi.

teachers in the school. The teachers jokingly commented that they are considering replacing Lenin's portrait with the one of the school principal, currently the only leader available at the school.

A NOTE ON REVERSE GENDER GAP IN MONGOLIA

Our Mongolian colleagues and friends couldn't resist teasing us for being fascinated with topics that were utterly unspectacular to them, such as the system of class monitors. Whenever we discussed senior officials or politicians they interjected jokingly, "he must have been an *angiin darga* [class monitor]!" These casual comments regularly ignited heated debate on why there are not more women in leading positions.

The *Country gender assessment Mongolia* (ADB and World Bank 2004) is not very flattering on the topic of women's representation in political decision-making positions. Mongolia scores far below the 30 percent benchmark set at the 1995 UN Fourth World Conference on Women. The most dramatic drop occurred at the first multiparty election in 1992 when the percentage of women elected into the National Parliament fell from 23 percent in 1990 to 3 percent in 1992. Since 1996 the National Parliament (elected in 1996, 2000, and 2004) has been composed of approximately 89 percent men and 11 percent women. The ratio is even more skewed when analyzing the gender composition of governors at the provincial *(aimag)*, rural district *(sum)*, city, and city-district levels. Of the 338 governors elected in 2001, only 3 percent (11 governors) were female (ADB and World Bank 2004: 4f.). The situation was quite different during the socialist period. For example, in 1980 women held slightly over 30 percent of the seats in Parliament (ADB 2004b: 93). To be fair, this imbalance is not a Mongolian but rather a universal problem. What is unique to Mongolia, however, is that women are excluded from the political realm *despite* being more educated than men.

The topic of reverse gender gap in education is omnipresent to the extent that all social ills are attributed to the "overeducation" of Mongolian women. From the inability to find a suitable husband, domestic violence, male drinking problem to male unemployment, the education of women is a scapegoat.[7] International researchers have discovered this trope, and they use Mongolia as a case study to understand why women at all levels of the educational system significantly outperform males. It is also a recurring theme in politics and is periodically stirred up in election years.

Female students are not only more likely to be elected as leaders in school, but they also stay longer. The figures for enrollment, non-enrollment, and dropouts speak for themselves: At all levels of the educational system females have a larger presence than males. The gender gap starts to widen dramatically after primary school. In 2000, the gap between the gross enrollment rate (GER) for females and the GER for males was 11.8 percent. This difference increased to 18.9 percent in the last two school grades (grades 9 and 10), favoring female enrollment. The trend continued for colleges and universities. Almost half of all female 18–22-year-olds (44.6 percent) study in colleges or universities, as opposed to only one-quarter of all males in the same age group (25.5 percent) (Steiner-Khamsi and Nguyen 2001: 38). At the tertiary level, the reverse gender gap has widened considerably. In 1990 the gap between females and males was 6.1 percent, as opposed to 19.1 percent 10 years later.

When we breakdown the total student population in tertiary education by gender, the disparities are immediately apparent. According to the 2002 UNIFEM/UNDP study, female students comprise 73.3 percent in tertiary diploma studies, 62.7 percent in Bachelors degree studies, and 65 percent in Masters degree studies (UNIFEM and UNDP 2002: 107). The underrepresentation of males in colleges and universities corresponds to international trends toward greater female enrollment at the undergraduate level. In Mongolia however it is cause for alarm, because it is linked with the high ratio of males who never attained the entry qualifications for higher education, either because they never enrolled in school or dropped out (see chapter 9).

Non-enrollment and school dropout is a serious problem for the government of a country that until 1990 had one of the highest enrollment rates for basic education worldwide. The gender imbalance is particularly high in rural areas. For years we heard and read about three popular explanations for male non-enrollment and dropouts in rural areas—male child labor, patrilocal residential patterns, and the rising private cost of education. The first is explained by the fact that boys are believed to be better equipped physically for herding in harsh weather conditions. The second is because women are married into the household of their husbands, and it is believed that educated women have a better chance than uneducated women of being treated respectfully by husbands and in-laws. Education is seen as a self-confidence booster enabling women to set boundaries, avert exploitation and, if necessary, to "talk back" to their husbands. The third reason—the private cost of education—functions as a filter for the first two. If poor herder parents are forced to make a decision concerning which of their children to enroll in school, they give preferential treatment to their daughters for the two reasons stated earlier.

A fourth explanation appeared in 2004: The lack of male role models in schools. The *Country gender assessment Mongolia* highlights the gender imbalance of school staff (ADB and World Bank 2004: 47). Whereas the teaching staff is predominately female,[8] school principals are almost exclusively male. The gender assessment report is comprehensive and well substantiated with data. It makes a brief mention of meetings with individuals in the educational sector who

> have identified the need to challenge these stereotypes by encouraging more men to work in primary and secondary education providing varied role models for young boys. (ADB and World Bank 2004: 47)

The author of the draft report, Helen T. Thomas, extends upon this to say: ". . . and to encourage more women to take decision-making responsibilities" (ADB and World Bank 2004: 47).

Mongolian policy makers, however, chose to adopt only the first half of the recommendation—to increase positive role models for young boys—and disregarded the second. Rather than encouraging a gender balance in staff at all levels, the female teacher was identified as the source of the reverse gender gap problem. Feeding into the general hysteria about strong Mongolian women—mothers, sisters, daughters, wives, and now teachers—producing weak men, action programs were hastily developed to bolster self-confidence in males. One of the programs, widely advertised during the 2004 election, was to introduce quotas for teacher education applicants favoring males over females. Using quotas as a policy tool is not new—in fact the former socialist government widely employed quotas for effective manpower planning—but it is unique to pay so little attention to the glass ceiling for women in the economic and political arenas. Statistically, women are more educated than men in Mongolia. This simple fact, however, does not make them more eligible to assume positions of leadership; in fact, just the opposite.

In Need of a Multi-Institutionalist Theory

In this chapter, more so than in others, we have drawn attention to cultural differences. It is not easy to offer cultural interpretations and refrain from oversimplification and essentializing. As important as it is to distinguish between class, race, and gender in the United States, it is essential to distinguish between rural and urban differences in Mongolia. For example, not talking back to the teacher is enforced as much in poor inner-city schools in the United States as it is in schools throughout Mongolia. Nevertheless, there is a shared understanding in each cultural context of how teachers and students are supposed to behave.

World-systems theory, especially in its expanded version, is useful for explaining globalization in Mongolia. The original concept of world-systems theory only considered economic factors (Wallerstein 1974), and distinguished between countries defined in economic terms as "core" and "periphery." Variations on this theme were later included (Chase-Dunn and Hall 1997). The original assumption was that the position of a country in the world-economy determines how individuals in that country behave, think, and feel. At the same time, the theory was attuned to intra-country differences, and explained why elites in peripheral economies identify with the metropolitan mind in core and sub-core countries, and thereby become local agents for the capitalist world order at home. For those who study postsocialist countries, the narrow concept of world-systems is odd as over 30 countries only joined the (capitalist) world-economy in the early 1990s. In previous chapters, we proposed to broaden the definition and include a political dimension that acknowledges the various political "spaces" that countries inhabit. Such a distinction proves essential for countries that have undergone major political changes, and have as a result had to reconfigure their alliances. After 1990, Mongolia not only abandoned the internationalist (socialist) world-system, but the government was forced to profess it would leave behind beliefs, policies, and practices reminiscent of a political space that now no longer existed. In some instances, the Government of Mongolia only adopted the language of the new allies to affirm its membership, while on other occasions policies and practices from the new allies were selectively borrowed and implemented. In this chapter, we propose further expanding the world-systems concept to consider the existence of separate pedagogical world-systems. This application is diametrically opposed to neoinstitutionalist theory.

David Baker and Gerald LeTendre (2005) present an impressive body of empirical findings to support the neoinstitutionalist assertion that educational systems in different parts of the globe are converging toward a world model of education. To do so they extract the data from two international studies on mathematics and science education and investigate global trends in schooling. The first study, Third International Mathematics and Science Study (TIMSS), was conducted in 1994, and focused on students in grades 4, 8, and 12. Besides test results in mathematics and science education in 41 countries, the TIMSS data set also includes an enormous amount of general information on teaching, learning, family, and the school, as well as specific information on the mathematics and science curricula in these countries. Five years later, researchers in TIMSS-99 tested students and administered questionnaires to students, mathematics teachers, and principals in additional countries, bringing the total to 53 participating countries in one or both TIMSS studies. It is important to point out that these international comparative studies of the International Association for the Evaluation of Educational Achievement (IEA) pursue rigorous technical standards for data collection and analyses, and operate with representative sample sizes. For this reason such studies are elaborate and expensive, and by default they exclude countries that don't have either the funds or the human resources available to follow their procedures. As with other IEA studies, few low-income countries participated; only 2 Latin American countries (Chile and Columbia) and 3 African countries (Morocco, Tunisia, and South Africa) were part of TIMSS and TIMSS-99.

Despite the sample bias toward high-income countries, the analyses of Baker and LeTendre (2005) are remarkable in that they systematically apply an institutional perspective to understand cross-national trends in schooling. In opposition to the view that schools as institutions are relegated to a state apparatus that merely reproduces inequalities to preserve the social structure, they see schools as the primary sites where social change is both mirrored and carried out. Moreover, social and institutional changes at the national level are reinforced at the global level, rendering local, regional, and national influences secondary. Gender differences in eighth grade mathematics test results, for example, have almost vanished in most of the educational systems that they studied. To use another example, the belief in education is universally embraced to the extent that parents, worldwide, are willing to invest in private tutoring and alternative forms of schooling. When access to quality education is at stake or concerns arise about whether teachers adequately prepare their children for standardized tests and university entry examinations, parents act to remedy the shortcomings of public schooling. The practice of private tutoring and the establishment of cram schools have become universal phenomena, and in some countries are quite extensive. In South Korea, Japan, South Africa, the Philippines, and Columbia three-quarters of the seventh and eighth grade students participate weekly in shadow education activities after school. However in other systems, for example in Scandinavia, the trend is not pronounced, because individual tutoring of students is considered part of the regular curriculum. The rapid and explosive growth of shadow educational systems is fueled by the high regard for schooling as a lasting "investment" in personal growth and professional fulfillment.

For our study on student-centered learning it is of particular relevance to review what Baker and LeTendre (2005) have to say on universal trends in pedagogy. In line with the neoinstitutionalist interpretation advanced by John Meyer, Francisco Ramirez, John Boli, George Thomas, Colette Chabbot, and others (see chapter 1), Baker and LeTendre emphasize the global spread of modern schooling. As a result, schools are not only organized and governed in a similar manner, but beliefs held and practices enacted by teachers are akin in different national educational systems. Analogous to the argument put forward by Boli and Thomas (1999) whereby a world polity (global civil society) steers decision making at the level of national governments, Baker and LeTendre (2005) attempt to present how every school is shaped by an internationally shared understanding of the practice of schooling. They write:

> If current trends persist, what happens in a classroom in Seoul, Paris, Santiago, Cleveland, or Tunis will be remarkably similar, most likely even more so than now.... The globalization of curricula and its implementation in classrooms will exert a soft but steady pull on nations towards a world norm, to the point where little variation in curricula exists across nations. What differences remain will be mostly across schools within nations for intentional reasons and some idiosyncratic variation introduced by teachers. (Baker and LeTendre 2005: 177)

Precisely because they see the various "grammar[s] of schooling" (Tyack and Cuban 1995) becoming gradually replaced by a universal, singular grammar of schooling, Baker and LeTendre (2005: 9) suggest that we conceive of schools as a "world institution" in which the same norms for modernity are reproduced and reinforced. The school homogenizes national differences in all aspects of instruction, including pedagogy. Based on the TIMSS video-study of German, Japanese, and American classrooms[9], Baker and LeTendre make the bold claim that national scripts of teaching are becoming increasingly irrelevant. Although they find the German, Japanese and American influence on lessons difficult to overlook, they advise the reader "not to focus too much on these differences, and not to assume that these patterns are stable across time" (Baker and LeTendre 2005: 113). What matters much more to them is the trend toward a "continued standardization of core teaching practices within academic subjects around the world" (Baker and LeTendre 2005: 115).

We would be hard-pressed to identify evidence that would suggest that mathematics is, or will be in the near future, taught the same way in Mongolia as it is in Germany, Japan, and the United States. This is not only because there are far fewer resources available to the teacher, but also because the teacher's professional identity and role in society are fundamentally different from what neoinstitutionalist theories are suggesting with regard to world culture. We are alluding here not to gradual but to fundamental differences in pedagogy. This is not to downplay internationally shared beliefs in "good" pedagogy. If the international organizations representing the world polity in Mongolia continue to fund in-service teacher training, it is likely that Mongolian teachers will buy into, and eventually believe in, the value of student-centered learning. What this imported pedagogy means to them, however, is completely different than what it means to teachers in Germany, Japan, or the United States. Beliefs about good teaching coalesce with beliefs about the good teacher and the good student, rendering the neoinstitutionalist assumption of standardized practices problematic. What converge internationally and eventually become standardized are pedagogical ideologies, not pedagogical practices.

Other scholars have also seriously questioned the claim that there is, or will be in the near future, only one (world) institution of schooling. William K. Cummings (2003) purposefully uses the plural "institutions" to denote the multiplicity of entities that disseminate modern conceptions of schooling around the globe. Applying an institutional perspective, Cummings finds six core educational patterns or six institutions of schooling. These institutions display distinct patterns of pedagogical scope, and as we show later, distinct conceptions of school and classroom technology, learning theory, administration, and administrative style. The following lists the period in which the six institutions were established, and presents the pedagogical scope pursued by each of the institutions of modern schooling (Cummings 2003: 35):

- Prussia, 1742–1820; whole person, many subjects, humanistic bias
- France, 1791–1870; cognitive growth in academic subjects, arts/science
- England, 1820–1904; academic subjects, civic and religious values, culture and curriculum
- United States of America, 1840–1910; cognitive development, civic values, social skills
- Japan, 1868–1890; whole person, wide range of subjects, moral values, physical and aesthetic skills
- Russia, 1917–1935; whole person, broad curriculum, technical bias.

As this list illustrates, not one but six institutions of schooling have spread across the globe. Cummings explains:

> [T]hese patterns were developed in the core nations of the world system and later diffused by their respective colonial and/or ideological systems. Thus, the French variant became influential in Africa, Indochina and Latin America; the English pattern was widely diffused through Asia and Africa; the American

pattern had some early influence in Asia and since the Second World War has had global influence; the Japanese pattern had a profound impact on Korea and Taiwan and more limited influence elsewhere; and the Russian socialist pattern influenced China, Eastern Europe, Cuba and many other developing countries. (Cummings 2003: 36–37)

Although we are in concert with Cummings' rejection of a single institution of modern schooling which supposedly has conquered the globe, we wonder why he selected those listed earlier, and why there are only six? Two points must be made before we proceed with presenting Cummings' contribution to institutional analyses of schooling. The first deals with his narrow conception of modern schooling, and the second addresses his choice to label the institutions after nations, rather than after global or regional school-reform movements.

Cummings prematurely dismisses Islamic education as "pre-modern" (Cummings 2003: 34) without further explanation. The project of modern schooling was first and foremost one of nation building, and as with Cummings' six other institutions, Muslim societies have also used schooling as an effective tool for horizontal and vertical integration in the modern nation-state. Horizontally, the same content is taught to different social groups in an attempt to create loyal citizens who imagine a common past and work toward a common future, as well as speak the same language and share the same values. Vertically, schools in Muslim countries also instill belief in social mobility whereby each individual is given—regardless of class, gender, ethnicity, and so on—an equal opportunity to improve her or his social status in society. The same controversies over the project of modernity apply for the Muslim institution of schooling: What is integration for some members of society entails exclusion, coercion, and discrimination for others. In other words, this project of modern schooling had the same positive *and* negative consequences, such as the oppression of linguistic minorities, differential treatment of social groups, gatekeeping functions, and other social stratification mechanisms that have preoccupied the sociology of education since its inception as a research field.

Second, Cummings' "core educational patterns" (Cummings 2003: 35) are too similar to earlier attempts in comparative education to determine the "national character" of each educational system.[10] They evoke unnecessarily negative associations with early twentieth-century historical functionalism in the field (see Pollack 1993). In contrast to the early comparative educational researchers in the United States (all of whom were historians), Cummings' occupation with core patterns acknowledges not only variations over time, but also differences within societies. Nevertheless, the national labels are misleading given that several institutions of schooling were in place in one and the same country. In the United States for example, until 1954 there were two separate institutions of schooling in place, one for whites and immigrants, and another for blacks and native Americans (Anderson 1988). As a result, the U.S. institution of modern schooling bifurcated into one that upholds beliefs in progressive education (later most commonly associated with John Dewey) and another committed to a racist notion of adapted education, manifested in the segregated South (Tuskegee and Hampton). This second institution was then transferred to the British colonial empire (Steiner-Khamsi and Quist 2000). Thus, if we were to take on the monumental task of identifying different institutions of schooling, we would consider regional or reform movements, rather than "core educational patterns" of educational systems, as criteria for developing the "ideal-types" (Cummings 2003: 35) of modern schooling. Despite these criticisms, Cummings' main point concerning the existence of multiple institutions of modern schooling is well taken.

In Table 6.1 we further extract four features of modern institutions of schooling that Cummings describes in greater detail (Cummings 2003: 35). We restrict our analysis to the institutions that Cummings associates with the United States, Japan, and Russia because traces of these three, or variants thereof, are discernable in Mongolia.

Striking in Table 6.1 is the different emphasis placed on individuals in the three institutions presented by Cummings (2003). Arguably, the concept of student-centered learning draws on the characteristics of "continuous development of the individual" or "individualized courses and instruction," attributed in Table 6.1 to the U.S. grammar of schooling. In contrast, other systems, including the Mongolian institution of schooling, place much more value on teacher-centered instruction and group orientation.

Applying a multi-institutional perspective is essential to our interpretation of student-centered learning in Mongolia. In chapter 2, we presented the different pedagogical world-systems that Mongolia inhabited, at times concurrently and at times subsequently, over a period of three centuries. Our historical analyses of educational import in Mongolia account for regional transactions and are therefore more specific than Cummings' global overview. Four of the five periods of educational import are analyzed in chapter 2: (1) enlightenment, (2) colonization, (3) nation building, and (4) universal access to education. Each period was distinct with regard to educational core objectives. Additionally, the reference societies varied across

Table 6.1 Core Educational Patterns in the United States, Japan, and Russia

Core Patterns in Select Countries	United States	Japan	Russia
Ideal	Continous development of the individual	Competent contribution to the group	Socialist achievement
School and classroom technology	Individualized courses and instruction	Teacher-centered, groups, school as unit	Collective learning
Learning theory	Aptitude and growth	Effort	Interactive
Administrative style	Management	Cooperation	Collective control

Source: Cummings 2003: 5.

time, and included Tibet, Manchuria, European countries and, during the final period, the Soviet Union. The other chapters in this book deal with the fifth period, beginning in 1990, when policy makers in Mongolia selectively borrowed traveling reforms designed in First World countries which were then, with funding from international organizations, disseminated to low-income countries. Despite the relatively brief description of the four earlier periods, the regional influences on educational development in Mongolia are essential for understanding the genesis of the Mongolian institution of modern schooling. Obviously, there was already an institution of modern schooling in place by the time structural adjustment policies and other educational reforms were transferred to Mongolia. Starting in 1990, the existing institution was not simply replaced by traveling policies and practices that had landed in Mongolia, but rather global reforms were interpreted and adapted to suit the Mongolian institution of schooling. In the same vein, the imported pedagogy of student-centered learning was Mongolized in ways that reflect the hierarchical social structure in the Mongolian institution of schooling.

7

Outcomes-Based Education: Banking on Policy Import

When outcomes-based education (OBE) first surfaced in Mongolia in 2003, observers noted that professional accountability had finally made it into schools. A new vocabulary accompanied the reform: benchmarking, scorecards, and outcomes-contracts *(ür düngiin geree)*. There was no doubt that OBE had been inspired by New Zealand's world-renowned model of curriculum reform. For years, Mongolian government officials were sent on study visits to New Zealand to learn about the New Public Management (NPM) model, more commonly known as New Accountability or New Contractualism. The reform model of "government by contract" (Schick 1998: 124) allegedly replaced interpersonal and informal agreements with measurable performance agreements or contracts.

In this chapter we address specific questions that globalization studies invoke. We utilize David Phillips's theory of "policy attraction" (Phillips 2004) to examine the context that initially triggered the import, and subsequently shaped the reinterpretation of OBE in Mongolia. This theory seeks to explain the sustained interest of policy analysts of one educational system in the educational provisions, reform strategies, and institutional features of another. Phillips observed British interest in German educational provision over a sustained period of time, and found that an educational system may be an object of attraction for a number of different reasons. Since the early nineteenth century, the British have been focused on the nature of schooling in Germany, styles of teaching and learning, the structure of vocational and technical education, and the German model of a modern university. Phillips's framework is cognizant of different policy contexts, and therefore relevant to our study on the OBE import in Mongolia. However unlike Phillips's longitudinal study, which seeks to understand why British policy makers were attracted to a particular aspect of the German educational system at a given moment, our study does not emphasize changes in policy attraction over time. Instead, we analyze OBE as a reform package in order to identify reasons why specific aspects of it resonated in Mongolia. To reiterate the point made in earlier chapters, borrowing is neither copying nor wholesale transfer from one context to another: it is always selective. Asking which aspects of a school reform package were particularly attractive to borrowers enables us to make the case that global reforms resonate for different reasons in various cultural contexts.

The question of *what* exactly was borrowed from the New Zealand OBE is important given that any major reform is an octopus with many arms. The OBE package extends into reforms that affect curriculum, monitoring of teachers, student assessment, teacher salary schemes, public accountability for the quality of schools, and, in some countries, school choice. When we studied how Mongolian teachers perceive OBE,[1] we found only two practices attributed to it: monitoring of teachers and performance-based teacher salaries. Other aspects, such as student assessment reform, curriculum reform, and public accountability, were so disliked that our interviewees wondered how we could possibly associate them with OBE.

Standardized testing for Mongolian language and math was introduced in 1998. Standards-based curriculum reform was launched the same year and still in progress in 2003, but not linked to OBE because it had gone dormant for almost four years. Curriculum experts in Mongolia had been at a loss as to how to tackle the colossal task of merging subjects in the overcrowded curriculum, while consolidating standards for each grade level and subject. However, the need to devise standards regained importance in 2002 when preparations were made to extend the curriculum from ten years to eleven. Introducing OBE at a time when content-based curriculum reform was already well underway meant effectively leapfrogging the crucial stage of developing concrete standards and learning objectives for each grade and subject. Data-driven public accountability for school quality had also already been well established by the time OBE was introduced. In Mongolia it was taken to a scale inconceivable in other countries. The "quality" of each school in math and Mongolian language instruction, measured as the percentage of students who scored an A or B grade on standardized tests, was published in newspapers and displayed at the Education and Culture Departments in the provinces. The Education Evaluation Center, affiliated with the Ministry of Education, has also printed thick books on the "quality" (grades A, B) and "success" (grades A, B, C, D) of *each* math and Mongolian language teacher in the country. These percentages reflect the test results of students in grades 4, 8, and 10, disaggregated by teacher. Teachers who score well are not only more likely to receive bonuses at school, but are also sought after by parents and hired as private tutors.

Interestingly, the two features OBE is commonly associated with—teacher monitoring and performance-based salaries and bonuses—were also already in existence. The *why* question, therefore, is not trivial. The OBE was imported from abroad when similar teacher monitoring practices, underpinned with an elaborate system of performance-based salaries and bonuses, had already been in place in Mongolia for three decades. Moreover, the reform was introduced at a time when serious doubts concerning the effectiveness of an outcomes-based approach to educational reform were being articulated worldwide. Given this contextual background, it is puzzling that a seemingly unnecessary and outdated reform was imported. In the case study of voucher-based (non)-reform in teacher education we explored the political and economic reasons for the import of the voucher *idea*. In contrast, OBE was less appealing as a concept than it was as a practice with financial benefits.

Finally, *how* questions typically deal with process. In chapter 5, we presented several examples of structural adjustment reforms that completely replaced previous approaches to schooling (e.g., grade 1–4 schools in remote rural areas). In chapter 6 we focused on hybridization of educational import, using the example of student-centered learning. The same applies here. OBE in Mongolian schools has been locally adapted to the extent that one wonders whether outcomes-based education is an unfortunate choice of terminology given that there is only a slight resemblance between OBE in Mongolia and OBE in New Zealand. In this chapter we take our interpretations a step farther. We argue that replacement (chapter 5) and hybridization (chapter 6) only represent two possible impacts of educational import. A third—reinforcement of existing structures—must also be considered. As absurd as it may sound, OBE merely reinforced the elaborate system of teacher surveillance that had long been in existence in Mongolia.

The Global Career of a Reform

Timing of educational import is essential, and Mongolia was a late adopter of the New Zealand model of OBE. New Zealand's OBE shares features with curriculum reforms that took place in England, Australia, Canada, South Africa and, for a brief period, the United States.[2] The Ministry of Education of New Zealand explicitly acknowledges the kinship ties of its reform with the National Curriculum in England (established in 1988), the Australian and Canadian curriculum reforms as well as South Africa's Revised National Curriculum Statement (Ministry of Education [of New Zealand] 2001).

A few comments on the workings of the "New Zealand model" (Dale 2001) will help explain why OBE is commonly associated with a fundamentally new approach to curriculum reform. The *New Zealand curriculum framework* (Ministry of Education 1993) places the individual student and his/her learning outcomes at the center of all teaching, and dissociates them from the content taught in a specific grade. In many countries where the New Zealand reform model was adopted, OBE requires that teachers establish benchmarks for each individual student. At the end of each grade, and in some countries throughout the year, the student's performance is regularly assessed by tests to measure whether the benchmarks have been reached. In practice, the proliferation of standardized tests is but one impact of OBE reform. In addition, the benchmarks are noted in the teacher scorecards, or outcomes-contracts, and teachers are held accountable for the performance of their students. Since OBE purports to measure the precise performance of a teacher as reflected in the learning outcomes of students, it has been propagated as a tool for quality enhancement in education, and aptly referred to as the New Contractualism, or New Accountability. In many countries OBE was accompanied by the introduction of merit pay or bonuses for teachers that did well on teacher scorecards. Claims have been made by proponents that OBE, as opposed to content or input-based curricula, monitors the quality of education more effectively, and better responds to the desire for greater public accountability in education. Finally, in many low-income countries, OBE was introduced to counter the adverse effects of structural adjustment policies on the quality of teaching (see Ilon 1994).

Of all the early outcomes-based educational reforms, the South African "Curriculum 2005," implemented in 1998, has great appeal to comparative researchers because it is a case of early adoption in a development context. Curriculum 2005 was modeled after earlier OBE reforms of the late 1980s and early 1990s in New Zealand, England, Australia, and Canada. There is a vast literature documenting the import of OBE in South Africa (e.g., Jansen and Christie 1999; Spreen 2004; Brook Napier 2003; Harley and Wedekind 2004; Todd and Mason 2005). More recently, South African scholars have drawn attention to the fact that South African experts were not only early adopters, but have subsequently become active disseminators of OBE in the region of Southern Africa (Chisholm 2005). Examination of how the OBE reform was exported regionally sheds light on South-South transfer, a topic that is, along with East-East transfer, severely understudied in globalization research.

Outcomes-based education was by no means confined to the African continent. In fact, the explosive growth of OBE reforms in the 1990s makes it necessary to conceive of it as a global reform movement,[3] and analyze it as a reform epidemic that spread to places previously immune to international trends. Surprisingly, Switzerland, notorious for its parochialism and procrastination[4] in complying with international agreements in general, and adopting international trends in school reform in particular, was on the front line promoting reform imports from New Zealand in the mid-1990s. What started out as a broad public administrative reform tailored after New Zealand's New Public Management (NPM)—introducing lean and efficient management, reducing the state apparatus, abolishing the status of civil servants, and replacing tenure of civil servants with performance-based promotion and employment—soon became the guiding principle for a major outcomes-based school reform in the Canton of Zürich, Switzerland (Steiner-Khamsi 2002). At the end of the 1990s, the radical reforms in Zürich caught the attention of policy makers in other European countries also considering moves toward educational reforms that were more outcomes-based and market-driven. By the time the governments of Mongolia and other postsocialist countries like Kazakhstan and Kyrgyzstan (Steiner-Khamsi, Silova, and Johnson 2006) joined the chorus of quality monitoring and public accountability, ministries and teacher unions in other parts of the world had reached the point of weariness with outcomes-based reforms.

Several ministries were requesting an evaluation of outcomes from the past decade, and in some instances reverted to more content- and standards-based curriculum reforms (see Donnelly 2002). In New Zealand, for example, the Ministry of Education mandated a review of the school curriculum in order to take "stock of what the curriculum reforms of the last ten years have meant" (Ministry of Education [of New Zealand] 2001). The New Zealand reforms of the early 1990s had also relied on school choice and the belief that the quality of education improves if schools have to compete for student enrollment. Proof that competition among schools increases the overall quality of education has yet to be found, and in fact many empirical studies of the New Zealand model, and other market- and choice-oriented educational reforms, point to the contrary (e.g., Ladd 1996; Ladd and Fiske 2003; Carnoy, Jacobsen, Mishel et al. 2005; O'Day 2002). In South Africa, the implementation of Curriculum 2005 has been postponed for another ten years because teachers and school administrators were at a loss as to how to translate outcomes-based or competence-based curriculum into practice (Brook Napier 2003). The Curriculum 2005 reform had to be relabeled, and is now referred to as the Revised National Curriculum Statement. Ironically, the period of great enthusiasm for OBE in Mongolia was concurrent with a nationwide strike by the National Union of Teachers in England against excessive high-stakes exams and teaching to the test (December 2003).[5]

It is typically difficult to map the trajectories of transplanted reforms that have had, as in the case of OBE, a "global career" (Dale 2001). The case of voucher-based teacher education reform in Mongolia is one example of ubiquitous policy transfer, and voucher reform traveled many routes before arriving in Mongolia. Reform models that go global are usually sufficiently vague to be embraced by many, yet precise enough to solicit compliance with "international standards" in educational reform. Vouchers are but one of many instances in which the spatial connotation of borrowing and lending research is rendered problematic, and we are urged to reflect on whether we should abandon mapping exercises in globalization research. Rarely can we identify one particular model that has been emulated; instead, selective borrowing from a large pool of voucher models, OBE models and so on is the rule. Carol Anne Spreen (2004), for example, found a global network at play in the import of OBE in South Africa. She traced the influence of experts from Scotland, Canada, the United States, Austria, and New Zealand on South African's Curriculum 2005, and documented study visits of South African policy makers to several countries that had already implemented OBE.

Contrary to all existing theories of research on policy borrowing and lending, ubiquitous policy borrowing did not apply for the import of OBE in Mongolia. In fact, OBE import in Mongolia seems to be the exception to the rule. The Mongolian Ministry of Education did not borrow eclectically from a large pool of existing OBE models, rather it selectively borrowed a specific one—New Zealand's—for a particular reason. New Zealand is in the same ADB region as Mongolia, and ADB funds were made available to transfer experiences from New Zealand—first NPM, and later OBE—to Mongolia.

Cross-Sectoral Transfer

The outcomes-based education reform in Mongolia was part of a larger public sector reform. The Public Sector Management and Finance Act, approved by the Parliament of Mongolia on June 27, 2002, advocated accountability and efficiency in the areas of governance and finance (Parliament of Mongolia 2002). The financial aspect of the law became known as the "re-centralization law" (Steiner-Khamsi and Stolpe 2004) because the newly established Single Treasury Fund was centrally organized. Hosted by the Ministry of Finance, the treasury replaced the previously horizontal structure of finance in which each ministry, including the Ministry of Education, was in charge of approving, disbursing, and overseeing its own budget. This new vertical approach to controlling public expenditures elevated the Ministry of Finance to an omnipotent agency, which all state institutions had to report to.

Accountability also implies dependence. Schools were made painfully aware of their reliance on the treasury at the end of December 2004 when, with only two-days advance notice, all school accounts were closed in a grand exercise of fiscal discipline, and all unused funds were returned to the Ministry of Finance.[6] In Mongolia, finance is the engine for any reform and, not surprisingly, the concept of accountability, permeating each section of the 2002 law, was linked to performance agreements and performance-based bonuses.[7]

The public sector management and finance reform was funded by a $25 million loan from the Asian Development Bank (ADB 2003). The first loan was approved in December 1999, and the second for $15.5 million, granted in October 2003, targeted accountability and efficiency in health, education, social welfare, and labor. In the late 1990s, New Zealand became the destination for policy pilgrimage. Every member of the Mongolian parliament, and all senior-level staff at the ministries, were sent on study tours to New Zealand. By the time all of this was taking place, however, critical observers had already published and widely disseminated their doubts about whether New Zealand's style of public management reform was applicable to developing countries (Bale and Dale 1998; Schick 1998).

As a late adopter of the new public management reform, the various ministries in Mongolia carried it out expeditiously.[8] In 2003, the Ministry of Education published a 319-page handbook on outcomes-based education with numerous examples of student benchmarks and teacher scorecards (MOECS 2003a). By the fall of 2004, this weighty tome had been distributed to all school administrations in the country who, in vain, browsed it to find solid criteria for evaluating educational outcomes.

The genealogy of OBE in Mongolia is important to consider when interpreting it as an example of cross-sectoral transfer. New Contractualism was the brainchild of Minister of Finance R. Tsagaan (Minister of Education since 2004), and his ministry pressured all public administration sectors, including the educational sector, to adopt the reform. In other words, all civil servants, including teachers, became subject to outcomes-contracts. Making civil servants in the educational sector adhere to the new public management reform was no small feat given that education constitutes the largest government sector (40 percent), followed by health (19.5 percent) and defense (13.1 percent) (ADB 2004a: 11). In contrast to other civil servants, however, teacher performance is typically measured by the learning outcomes of students. Although the monitoring experts at the Ministry of Education listed numerous examples of student assessments in the OBE handbook, teacher performance and student outcomes are not so easily linked. From the perspective of civil servants, New Public Management was a means to pump more money into the system. Unlike other countries where the system of performance-based salaries evokes anxieties about job loss, the fear is unsubstantiated in Mongolia. Labor laws guarantee job security, making it difficult to terminate a contract no matter how poorly an employee performs. Against this backdrop, civil servants including school employees, hoped for monetary rewards for scoring high in the outcomes-contracts. Getting punished for low performance was not among their concerns.

Discipline and Punish Teachers

In the Teachers as Parents study (Steiner-Khamsi, Tümendemberel, and Steiner 2005), we were surprised to find an elaborate system in place to discipline and punish teachers for their shortcomings. In provincial schools, teachers are held personally accountable by parents and the school administration. This occurs not only if students do not excel academically in class, but also if they do not do their homework, take proper notes, engage in useful after-school activities, clean the classroom, or maintain hygiene. Teachers have ambivalent feelings about this treatment. On one hand, the high expectations mirror the honorable status of a teacher in Mongolian society. The title "teacher" (*bagsh*), is also still given to knowledgeable or wise individuals outside of the teaching profession. Most of the teachers we met in Mongolia felt they must live up to these expectations. On the other hand, they found some of the demands excessive, and were critical of how much responsibility a teacher is expected to take for students who "don't listen in class," or "don't do the reading and writing." "Parents come to school and yell at us teachers if their child performs poorly," was a recurring complaint in rural schools. Voluntarily tutoring students after school to avoid being humiliated by parents, or suffering a salary reduction, is an unfortunate aspect of the profession for teachers in Mongolia. In cities, the status of teachers has declined rapidly since the dramatic changes of 1990, and public shaming and humiliation of teachers has been replaced with institutional technologies. OBE is among the technologies that have reinforced preexisting teacher surveillance practices.

The coexistence of several performance- and outcomes-based monitoring systems has significantly increased the bureaucratization of the teaching profession. Drowned in paper work, teachers must submit their daily notes on students as well as monthly and semiannual self-evaluations to the school administration. In the absence of clear evaluation criteria, the excessive reporting is nothing but useless busy work and is essentially an indicator of whether a teacher is doing what they are told. The leapfrogging—moving directly to OBE without a prior establishment of standards—has left school administrators and teachers wondering how, and against which benchmarks, they should evaluate outcomes. Because there are no concrete standards against which teacher performance could be determined, the school administrators assess the quality of teaching arbitrarily, feeding widespread suspicion of nepotism and favoritism. Even worse, consistent with OBE reforms in many other countries of the region, there are no budgetary implications for better outcomes. The school administration is trapped in a zero-sum game where to reward a few high-performing teachers, it must deduct income from others.

The system of personal accountability relies on myriad regulations to keep teachers in line. The Mongolian performance-based teacher salary scheme, in place in various manifestations for the past 30 years, is an important factor in the discipline and punishment of teachers. The full income of Mongolian teachers has traditionally consisted of base salary, salary supplements, and bonuses. However today base salary only constitutes approximately 57 percent of total income, while salary supplements and bonuses represent a sizeable share.

The *base salary* ranges from between $45 and $55 per month, depending on the rank of the teacher. The salary scheme differentiates four ranks: regular teacher, methodologist, lead teacher, and advisor. Promotion to a higher rank entails an increase of 5–25 percent of the base salary.[9] The promotion criteria vary slightly for the different ranks, but they all include leadership skills (number of teachers that have mentored or trained), ethics of the teacher (appearance, communication skills, personal behavior), grades of students, and awards from "olympiads" and competitions. Of all these criteria, winning at olympiads carries the greatest weight, making all other measures inconsequential. Furthermore, if a teacher wins at a high-level olympiad (provincial or national level), they are able to skip ranks and get promoted directly to lead teacher or methodologist. Interestingly, both the promotion and salary increase also go into effect if a student wins at an olympiad or competition. The assumption is that the teacher supported and promoted the award-winning student, and therefore deserves to be rewarded in return. Critics of olympiads point to the detrimental effects of linking teacher salaries and bonuses to student outcomes in olympiads. This practice is said to encourage teachers to focus solely on a few promising students, coach them for olympiads, and neglect the rest of the class.

The system of olympiads is a relic of the socialist past that remains popular among teachers, students, and parents. The practice of socialist competition was ubiquitous, and embraced all groups (teachers, students, workers, herders, mothers, etc.) and all state institutions (factories, agricultural collectives, animal husbandry collectives, government offices). The competitions were conducted at each administrative level—municipality, district, provincial, and national—leading to a whole host of awards and insignia. It is likely that each and every citizen won a socialist competition for something: being the best worker, the best student in mathematics, or the best stamp collector, to list a few examples. The importance of olympiads and other performance-based promotion criteria was reaffirmed in government regulations of the postsocialist era, notably in 1995 and 2004. A common reaction among the interviewed teachers toward OBE, and the teacher scorecards in particular, was that the emphasis placed on outcomes and performance was socialist competition in disguise. In addition, socialist-oriented teachers viewed OBE as an egotistical version of socialist competition in that it advocates competition without a sense of social responsibility for the group or collective.

The *salary supplement* constitutes regular monthly income and is given for numerous tasks, ranging from teaching a class for gifted students (20 percent supplement to the base salary), serving as a class teacher or grading student notebooks. Students are expected to take notes on what the teacher says and consolidate that knowledge in their assignments. Every teacher—including physical education, art, and music teachers—is rewarded on the basis of having demonstrated, through their students, that they are knowledgeable.

In contrast to salary supplements, *bonuses* are usually onetime awards given throughout the school year for special accomplishments by teachers. In the absence of clear evaluation criteria however, they are determined arbitrarily by the principal, the education manager (assistant principal), and the social worker,[10] who together constitute the school administration. The teacher scorecards, introduced in the wake of the OBE reform, fall into this category for two reasons: A high score (i.e., 60 percent or more of all eligible points) calls for a bonus. However the ambiguity of the evaluation criteria entails that the school administration values the teacher for reasons unknown to all other teachers.

As common as it is to obtain a salary supplement or bonus, it is also common to lose it or to have one's salary, supplement, or bonuses deducted. Salary deductions are serious and only made for teacher absences, tardiness, or drunkenness.[11] In contrast, deductions of salary supplements are very common. The school administration establishes its authority with teachers by constantly threatening to reduce their income, creating an atmosphere of intimidation and obedience in the school.

It is a striking feature of the Mongolian educational system that the laws and regulations are formulated meticulously, and the sanctions for not obeying them are as well. Along with the host of regulations imposed by the district, provincial, and central education authorities, each school also develops its own additional policies. In a school in the Bayangol city-district of Ulaanbaatar, for example, the supplement for grading student notebooks is only given if the teacher successfully enforces the following seven requirements of student notebooks:

1. Full name and address on notebook cover, written in proper handwriting
2. Tidy notebook cover (i.e., not spoiled and not ripped)
3. No crossed out or corrected words

4. Legible and neat handwriting
5. Complete and correct notes on the teacher's lessons
6. No mixing of ink in the same notebook[12]
7. Evidence that the teacher actually checked and corrected the student notebook. (Bayangol 2004a)

Noncompliance with any one of these seven requirements results in a supplement deduction.

Another example of this punitive system is the regulation of class teachers. In the same school in the Bayangol city-district of Ulaanbaatar, seven criteria must be fulfilled for class teachers to receive the full supplement. The criteria are equally weighted, each carrying a maximum of ten points:

1. Cleanliness of classroom: 10 points
2. Discipline of the class: 10 points
3. Clothes and appearance of students: 10 points
4. Condition of class furniture and equipment: 10 points
5. Attendance of students: 10 points
6. Making use of the class bulletin board: 10 points
7. Accomplishment of given duties and responsibilities: 10 points. (Bayangol 2004b)

The policy also clearly lays out the infractions that result in a reduction of the class teacher's supplement:

- For every student who is not disciplined: 5 point deduction
- For loss or damage of classroom equipment and furniture, each instance: 5 point deduction
- For not updating the class billboard (class newspaper, posters, etc.): 5 point deduction
- For every student who comes late or misses class: 1 point deduction (Bayangol 2004b)

It is important to bear in mind that the regulations for the class teacher supplement are in effect for all levels, from preschool to higher education. Furthermore, the class teacher is held accountable if a student acts out or is "not disciplined" on the school premises, before class, after class, or even on the weekend. In this case, the assumption is that the class teacher clearly did not serve as a role model, and did not mold the student into a respectful and respectable human being.

As previously mentioned, teachers are only eligible for a full salary supplement if they score 70 points, or more precisely—as the following listing indicates—63–70 points (90 percent–100 percent). The full supplement is T 8,700 ($7.70) per month, with a guaranteed minimum amount of T 2,500 ($2.20). The school administration and teachers of the Bayangol city-district school agreed at the beginning of the 2004–2005 school year to use the following award scheme for class teachers:

90%–100% [of the total of 70 points]:	T 8,700
81%–89%	T 7,400
70%–80%	T 6,840
60%–69%	T 5,600
50%–59%	T 4,750
41%–49%	T 3,910
40% and less	T 2,500

(Bayangol 2004b)

The award scheme for class teachers serves as scaffolding for the figment of data-driven performance pay in three distinct ways: First, it evokes scientific rationality by presenting exact figures that are treated as quintessential authoritative sources, rendering administrative decisions irrevocable. Second, the rules are made transparent suggesting collective and democratic decision making. Third, the small performance intervals used in the award perpetuate the illusion of a sophisticated apparatus that is capable of assessing teachers in exhaustive detail. The 7 criteria, the 4 types of infractions, and the award scheme for class teachers, all illustrate a culture of teacher surveillance that has flourished over the past 3 decades.

To offset the frequent loss in salary supplement, the class teacher puts the class monitor in charge of collecting fines from students who come late or miss class, and requests "donations" from parents to fix or replace old desks, chairs, and blackboards. This way, the deductions made by the school administrators are partially compensated with private contributions from parents. The parents feel obliged to comply with this practice but resent being constantly asked for financial contributions.[13]

THE OBE TEACHER SCORECARDS

OBE was introduced in the 2003–2004 school year, becoming part of the culture of tight surveillance with salaries tied to performance. Teacher scorecards are the most visible marker of OBE, signaling a distinction between what had already been in place with regard to quality monitoring and the newly added system of teacher surveillance. In Mongolian schools, the teacher scorecard is a sheet of paper entitled "outcomes contract," typically translated and written with quotation marks to denote the specific usage of this uncommon term: "*ür düngiin geree.*" Each school develops its own outcomes-contract, but most of the contracts between teachers and the school administration are strikingly similar. In many schools the OBE contract is used alongside other teacher-performance measures, aligned with past experience. During the socialist period, teachers had to memorize decisions of the Party Congress related to education, just as today they must memorize the laws, guidelines, and regulations issued by the Ministry of Education. In a

Table 7.1 Outcomes Contract, Example Töv Province, " 'Outcomes Contract' between Teacher . . . and the School Administration"

No.	Outcomes	Teacher's Comments	Evaluation	
			Self	Administration
1	Class Management (0–10)			
2	Lesson Planning (0–15)			
3	Student Development (0–15)			
4	Teacher's Note Taking (0–10)			
5	Teaching Skills (0–10)			
6	Teacher's Creative Work (0–10)			
7	Professional Development (0–10)			
8	Time Management and Task Completion (0–10)			
9	Teacher Morality and Responsibility (0–10)			
10	Maintenance of Property and Cleaning of Classroom (0–5)			
		Total Points		

Names and signatures of school administration: Name and signature of the teacher:

_____   _____

Principal   Teacher

_____

Education Manager

_____

Social Worker

Source: Zuunmod 2004.

school in the Övörkhangai province, for example, teachers reported that the OBE contract only accounts for 70 percent of the score; the remaining 30 percent consists of teachers' knowledge of educational law.

Table 7.1 provides an example of an outcomes contract in Zuunmod *sum*, Töv province (Zuunmod 2004). It includes ten outcomes that must be evaluated by the teachers themselves and then passed on to the school administration. Each of the 10 outcome categories carries 0–5, 0–10, or 0–15 points. Zero points implies that the expected outcome was not achieved.

Arguably, the imported OBE reform has been locally adapted, or "Mongolized," in substantial ways. The basic tenets of the original New Zealand model are barely discernible. For example, indicators measuring learning and teaching outcomes have less weight than indicators reflecting how well the teacher fulfills administrative tasks and interacts with others in school. Three of the performance criteria address administrative functions of the teaching profession in Mongolia. "Official documents and notes" measures the practice of taking excessive notes on teaching resources used in class, student registration, and reports periodically submitted to the school administration. "Time management and task completion" indicates how often the teacher was late or absent from class, and whether she has carried out the tasks given by the school administration in a timely manner. "Maintenance of property and cleaning of classroom" is regularly monitored by the education manager (assistant principal). Finally, the "teacher morality and responsibility" category includes three dimensions: the teacher's communication style toward students, parents, and school staff (abstention from verbal and physical abuse), the teacher's appearance and habits (professional dress code and sobriety), and the teacher's respect and subservience toward the school administration.

Of the ten performance criteria, only two deal explicitly with students—class management and student development—both of which are evaluated by the teacher.[14] The indicators of teaching quality ("lesson planning," "teaching skills," "teacher's creative work") are also self-reported and rarely observed by the school administration. For the sake of convenience, the education manager merely evaluates the factual aspects of teaching quality: the thickness of the notebook in which the topic for each lesson is to be listed ("lesson planning"), and the number of teaching resources or booklets the teacher produced in the past school year ("teacher's creative work").

The teacher must fill out the self-evaluation form each month and hand it in to the education manager. According to the teachers we interviewed (Steiner-Khamsi, Tümendemberel, and Steiner 2004), the education manager quickly reviews these sheets, checks them off, and places them in his drawer. It is only toward the end of the fiscal year (December), once the school accountant informs the school administration about the savings made from school maintenance, repair, parental "donations," or salary and supplement deductions, that these teacher scorecards resurface from the education manager's files. It is at this critical moment that teacher performance on the OBE contract and other accomplishments, notably at olympiads and competitions, are reviewed for bonuses. For example, at the school in Zuunmod, Töv province, teacher scorecards are reviewed semiannually. At the end of the fiscal year, the school succeeded in making considerable savings, and rewarded its teachers with a bonus if they obtained 60 percent or more of the available points on the teacher scorecard. Typically, the administration must reinvest its savings into the general school budget and is not in a position to reward its employees with extra funds. Instead, it applies the zero-sum approach mentioned earlier: to reward a few school staff with bonuses for scoring high on the outcomes-contracts, the administration makes deductions from salaries, salary supplements, or bonuses allocated to other staff.

The New Zealand Model: A Lucrative Reform

It has been repeatedly argued that the system of performance-based salaries is not new to the Mongolian education sector. In fact, there was already an elaborate and bureaucratic mega-structure of surveillance in place, underpinned by myriad policy documents and laws regulating, administering, and legitimizing the system in minute detail. The purpose has always been to control teachers by rewarding and punishing them with either salary supplements/bonuses or deductions. What is new, however, is the fact that teachers now have to self-evaluate. From a Foucauldian perspective, one might propose that the OBE reform institutionalized a modern technology of surveillance that demands insight, remorse, and continued self-betterment (Foucault 1995; Popkewitz 1998; Rose 1998). In practice, however, the outcomes contracts encompass a wide array of objectives, many of them matching the concerns of school administrators, with little relevance for learning outcomes.

Data-driven accountability or results-oriented performance in a setting that is regulated by interpersonal networks, imbued with expectations of reciprocity, and devoid of clear performance standards, is predestined to advance favoritism and nepotism, and discourage teachers who hold high quality standards.[15] A case in point is our finding (Steiner-Khamsi and Gerelmaa 2005) that teachers neither knew why exactly they received a bonus, nor why their salary supplement was deducted. Those who benefited from OBE viewed it as another accountability system that helps them to boost their income. At the same time, they felt embarrassed that they were given preferential treatment for no obvious reason, and at the expense of their colleagues in the school. Those who were left empty-handed at the end of the outcomes contract ridiculed the teacher scorecards for relying exclusively on self-evaluation and self-reported accomplishments.

In this case we were only able to draw from two small empirical studies: the Teacher as Parents study (Steiner-Khamsi, Tümendemberel, and Steiner 2004) conducted in 4 schools, and the pilot study of the Public Expenditure Tracking Survey (PETS) administered in 6 schools (Steiner-Khamsi and Gerelmaa 2005) in 2005. The Public Expenditure Tracking Survey of the World Bank is likely to yield exact figures on the full composition of teacher income.[16] Despite the explorative nature of our two studies, the perception of OBE as a means to generate additional income is clear. Both teachers and government officials expected financial gains from OBE.

Since the socialist period, money has continually been the incentive for reform in Mongolia. Interestingly, some of the new imported reforms are exact replicas of what was already in place. As such, OBE in Mongolia qualifies as a quasi-reform used as a "flag of convenience" (Lynch 1998: 9) to secure international funding. Linguistic nuances are important. Rather than claiming that the reform was funded with the support of large loans from the Asian Development Bank, it is more accurate to state that the Ministry of Finance was given large loans *for* implementing the New Zealand management and finance system. As with other donor involvement in Mongolia, loans and grants are accompanied with reform packages. The reliance on external funding becomes immediately apparent when we consider the large number of "national programs" or "action plans" that are publicly announced but never implemented due to lack of funding. In contrast, by using the language of New Contractualism, funds were secured, and OBE could be added to the pile of already existing accountability systems.

Both the Mongolian government and the schools were able, quite literally, to bank upon the OBE reform. The dubious evaluation criteria for outcomes notwithstanding, a well-performing teacher in a Mongolian school can now accumulate bonuses from different accountability systems and considerably improve her low base salary. By adding one more monitoring policy (OBE) to the series of firmly established teacher surveillance technologies, the government was able to secure loans, and some teachers were able to acquire bonuses. Banking on policy import was an important reason why OBE was so appealing to government officials, school administrators, and teachers.

8

SPEAKING THE LANGUAGE OF THE
NEW ALLIES WITH THE
VOUCHER (NON-) REFORM

In 1998 the Ministry of Education introduced vouchers for its in-service training of teachers. The initiative was announced publicly as a fundamental reform that would replace the Soviet-based system, and herald a new age of democracy, free market economy, personal freedom, and individual rights. The Ministry claimed it would promote "lifelong learning," "decentralize" the system, break the "state monopoly," cater to the "individual needs" of teachers, advance "choice," and "enhance the quality" of teacher training by permitting "competition" from other training institutions. The reform further sought to demonstrate that Mongolia was not falling behind international standards in educational reform.

As with educational reforms worldwide, there was a large gap between what was initially announced in public, subsequently enacted on paper, and eventually implemented in practice. The vouchers have in fact ultimately been used as (unevenly distributed) registration forms. Contrary to what was announced earlier, there has been no sign of a demand-and-supply-driven training reform. It is hardly news that a reform may play out differently at the levels of "policy talk" (Tyack and Cuban 1995; Cuban 1998), enactment, and implementation. But the policy analysis presented here goes beyond mere description of these apparent differences to reveal the meanings and expectations various stakeholders attached to the reform. The study of voucher-based teacher education reform is especially intriguing, because it is a case of a policy import whose underlying concept (public choice) never made it into practice. As such, it is an example of discursive borrowing, or more succinctly, borrowing that is "phony" (Phillips 2004) or fake.

This case analysis draws from data we collected in School 2001 partner schools during the first three years of the voucher reform (1998–2001). As the largest in-service teacher training program in Mongolia, School 2001 established school-based and regional in-service training in 72 locations from 1998 to 2001. As we explain, NGOs had a vested interest in making the voucher-based reform work. It would have sustained their initiatives, and their certified trainers would have been reimbursed by state-issued vouchers rather than by their own funds. Using the large sample composed by the 72 partner schools of the School 2001 project, we explore how the voucher plan was implemented.

This sample is not representative of all schools. Each of the 72 schools saw themselves as part of a movement, and strongly advocated for any kind of in-service teacher training reform in Mongolia, with or without vouchers. This means that the respondents were perhaps more opinionated in their vision of the need for in-service training reform than teachers and principals from other schools might have otherwise been. The findings presented here draw from the qualitative data we gathered after numerous school visits and interviews with teachers (both individual interviews and focus group interviews), principals, and administrators in Ulaanbaatar and seven provinces during the period from 1998 to 2004.[1] We also interviewed subject matter "methodologists" who are formally in charge of in-service training at the province level.[2] In Ulaanbaatar, we spoke with all senior staff in charge of in-service training at the Ministry of Education, the State Pedagogical University, and the Mongol National University. Statistical evidence on in-service training (type of training, number of participants, content of training, and qualification of trainers) is available for the period 1998–2001, but it is not included in this book.[3]

This chapter presents thick descriptions of the voucher-based system. The system was perceived differently by the three principle professional groups involved in the reform: government officials, school administrators, and teachers. In doing so, we account for situated knowledge in these different professional groups in order to understand the specific reasons why a particular group either supported or opposed the voucher-based reform. Thus, the selection of quotes represents prototypical statements that a majority of group members made when they spoke about the voucher-based training reform.

The research questions for the voucher study evolved over time. We began by simply asking whether vouchers were actually used. Our research then developed into a series of follow-up questions regarding the features of the Mongolian model of vouchers. From 2001 onwards, the research focused on the reasons why the voucher system failed in Mongolia.

POLICY TALK ABOUT VOUCHERS

In Mongolian, the word "voucher" (*erkhiin bichig*) is a composite term describing a document that provides an individual with specific rights. The concept of vouchers was applied in the first two privatization laws (July and October 1991). Red and blue vouchers had been in circulation during this first wave of privatization, when shares of state and collective enterprises were distributed to individuals (Boone 1994: 350). All citizens born before June 1991 were eligible for vouchers provided they had the proper registration documents. Red vouchers were used for small-scale privatization (small shops, livestock, agricultural equipment, trucks, etc.), and blue vouchers for large-scale privatization (shares in large enterprises). Beginning in 1991, vouchers were used to inscribe property rights, evoking associations with a market economy (Schmidt 1995; Korsun and Murrell 1995).

In education the label—frequently referred to in English and written in Cyrillic (*vaucheriin sistem*)—raised all sorts of grandiose expectations. Shortly after their introduction, a rumor circulated that the Ministry of Education also planned to distribute vouchers to parents, extending choice from teacher in-service training to education in general. In informal settings skeptics raised their eyebrows and dared to question whether choice could ever successfully resonate in Mongolia, given that in the provinces the average distance between schools is 45 miles. A memorable conversation from that time occurred between two Mongolian education experts, questioning whether choice would benefit or harm the traditional nomadic lifestyle of the Mongolian herders. Many pundits viewed the government's enthusiasm for vouchers as an early morning mist that would eventually dissipate and reveal a plan to completely revamp Mongolia's educational system according to American expectations.

These fears were not entirely unfounded. The government did entertain the idea of introducing school choice. In 1997, the Ministry of Finance and UNDP commissioned a study to examine whether the introduction of vouchers would curb public expenditures for education while preserving access to schools in rural areas and enhancing the quality of education throughout the country. S. Lhagve, author of *Internal market in education*, attempts "to identify the specific problems which may be caused by [the] introduction of a financing system in education based on 'public choice' theory, particularly [the] 'voucher system' concept" (Lhagve 1997: 3). Reflecting on the educational reforms during the early transformation period (1991–1997), Lhagve reveals several "problems which were brought from the [socialist] past and contradict with an emerging market system" (Lhagve 1997: 6). He points out that the first step in transforming the centrally planned economy to a free market economy had already been taken, albeit with mixed outcomes. Education was decentralized, at least on paper.

Lhagve's study seeks to answer whether the second step is desirable: the move from a state-planned to a "public choice"-driven education initiative in the form of school vouchers. He provides a detailed account of all that went wrong in educational reform during the early structural adjustment period. The study develops a series of grim scenarios of what could happen were vouchers to be introduced: an acceleration of rural-urban migration leading to a decline in class sizes, or even a "closing effect" (Lhagve 1997: 15) in rural schools, which would then result in overcrowded classes, as well as bribery and corruption in well-performing urban and semi-urban schools. Lhagve's forceful warning against the introduction of school vouchers was well received by the government. Nevertheless, the underlying idea of "public choice" in education as a necessary next step toward creating a real free market economy was convincing, and has remained a recurring theme in education.

In March 1998, four months after *Internal market in education* was released, the introduction of a voucher-based reform was announced. Schools were exempt, but in-service teacher education was targeted. As the policy began to take shape, speculations about the far-reaching effects of the voucher system on both the education system and on the Mongolian way of life began to fade. It became clear that the reform was far more modest than expected.

The Enactment of the Voucher Policy

There are two official documents that deal specifically with the voucher-based reform. The policy itself was signed (March 24, 1998) by the finance and education ministers (Mongolia National Government 1998), and the "national program" was authorized (June 5, 2001) by the Prime Minister and the Minister of Education (Mongolia National Government 2001). All political parties in power expressed their commitment to vouchers. The coalition government of the Democratic Union initiated the reform in 1998, and the Mongolian People's Revolutionary Party (MPRP) government, reelected in a landslide victory in 2000, confirmed the decree issued under the previous government, adding only a few administrative details to the new one issued in 2001. In 2003, the MPRP-led Ministry of Education invited governmental and nongovernmental institutions to apply as providers of in-service training. International NGOs hesitated, however, because this would have entailed having their trainers certified and closely monitored by the Ministry of Education.

The first voucher decree (March 1998) meticulously outlines procedures for the management, distribution, and accounting of vouchers in the "voucher-based in-service training system for teachers and administrators." The policy distinguishes between "three types of training": workshops at the central and regional levels, and "independent learning." Each year the Ministry of Education and the Ministry of Finance determine the number of vouchers for the upcoming year, and specify how many are to be used for each of the three types of training. The policy document also stipulates that the Education and Culture Centers (education administration units in the provinces and in the cities) hand out vouchers directly to the trainees (Mongolia National Government 1998: sections 1.2, 1.3, 1.4).

The previous in-service training system was directly offered by the School of Education Development, the state-run in-service training agency. Under the new program, a wider range of state institutions were given permission to offer in-service training, provided they had received special permission or accreditation from the Ministry of Education. In the same year, an amendment to the Education and Higher Education Laws was passed, which listed teacher training as one of the new responsibilities of universities (see Innes-Brown 2001). The opening up of in-service training to different providers was an acknowledgment, unprecedented in the history of Mongolian in-service teacher education, that teachers have individual needs and interests and should therefore decide which courses and resources are most suitable for their work. As the following excerpt illustrates, the 1998 decree was steeped in the language of "choice":

> Teachers who receive vouchers will read the announcements of the training organizations, and then select the organization based on their own interests.

The Independent Learner, in turn, will produce a list of teaching resources that he/she wishes to purchase. He/she submits this list to the Education and Culture Center within one month of enrollment as an Independent Learner. (Mongolia National Government 1998: section 2.5)

The second voucher degree of 2001 preserved the core idea of choice. It further promoted an institutional linkage between preservice and in-service teacher training, advocated an additional type of training (school-based in-service training), and encouraged the participation of nongovernmental and international organizations, as well as the public sector as training providers. More so than in the first decree of 1998, the second decree of 2001 explicitly underscored the important role of teacher education reform for raising the quality of education in Mongolia. The decree begins by stating the following:

[T]he development of Mongolian society necessitates a deepening of educational reform in Mongolia to enhance the quality of education, to adjust education to the needs and interests of the citizens, to improve the content and the teaching methods used in pre-service teacher training, and to build an effective system of in-service training. (Mongolia National Government 2001: 1)

The visions inscribed in the introductory sections are arguably misleading, and they stand in stark contrast to the regulations meticulously outlined in the remainder of the policy documents. On one hand, the utopian language used in the preambles reflects the "age of the market" *(zakh zeeliin üye)* (Sneath 2002a: 195); but, when it comes to regulating the enactment of the policy, the documents merely reiterate what had already been in place since the 1990s. The encrypted messages in Mongolian policy documents require special skills to be deciphered. Once the reader moves beyond the layer of messianic prophecies—today evoking the age of the *true* market, and in the past, the age of *real* communism—the reality of a shortage economy surfaces. The fact that the budget for in-service teacher education was not increased but rather expected to be reduced, sent a signal. Unlike experiences in other countries, where the introduction of vouchers had been used to revamp the structure and provision of education, the Mongolian voucher experiment was not designed to trigger a fundamental reform. Also different from other countries (Carnoy and McEwan 2001; McEwan 2000), the introduction of vouchers in Mongolia did not lead to soaring public expenditures for education. Rather, the Mongolian reform was based on very limited financial resources and reflected the general cutbacks that the education sector had experienced throughout the 1990s.

Vouchers legitimized the practices that had gradually developed in the past decade: short in-service teacher education programs at the central level (in Ulaanbaatar) and regional levels (in the provinces). Additionally, they preserved the provision of Independent Learning that existed during the socialist period. By adjusting the value of the vouchers to the "type of training" and "location of the school," it further institutionalized inequality between the (expensive) central and (cheap) regional training that had evolved over the past few years. Explanation for the paucity of resources underwent an astonishing metamorphosis in the two policy documents. Rather than being associated with Ministry of Education cutbacks on teacher education, the lack of resources was attributed to the "age of the market," which after a period of hardship or transition, would eventually provide quality and effective teacher education for all. Deferring to a distant future has an established tradition in Mongolia, and is also discernible in the language of the voucher policy documents. In the past, it was the move from state socialism to communism; in the 1990s, it was a move from the "first step" (free market economy) to the "second step" (public choice) (Lhagve 1997: 6).

Voucher Implementation

When the Ministry of Education and the Ministry of Finance formulated the voucher-based system on paper, they downplayed the original concept of individual choice, encouragement of private sector involvement, and diversity of courses. Subsequently, the staff in the Ministry of Education and the provincial education authorities in the Education and Culture Centers (ECC), in charge of implementing the voucher system, further watered down the original intent. By the time the voucher idea actually reached teachers, it had little to do with choice. At the policy implementation level, the voucher-based system was soon transformed into a nepotistic system that benefited ECC directors, methodologists, and school administrators, leaving the mass of teachers untrained.

Vouchers were a public issue between 1998 and 2002, but only senior administrators were familiar with the content of the voucher decrees. Our interviewees had a great interest in speaking with us about the reform and hoped to learn more about the idea of choice and the logistics of the voucher-based system. These educators and administrators expressed different reasons for dismissing the current system as dysfunctional. In the following sections, we present the experiences of each professional group with the voucher-based reform.

Teachers

"What vouchers?" was a common reaction among teachers. Over a period of six years, not one of them, and only a few of the assistant principals, ever held a voucher in their hands. They had heard of vouchers, but thought they were not intended for regular teachers. Junior teachers tended to believe that the vouchers were to be spent on the training of principals, assistant principals, and methodologists. They were given the impression that vouchers were intended to train trainers, especially methodologists, who would then return to the provinces to train teachers. Several senior teachers attended such workshops and complained about both the lecture style and the abstract nature of the content. One informant claimed: "All they are teaching is about core principals. First it was about communist core principals, and today it is about international ideas of how society functions, and what role education should have." Senior teachers wished that they still had the old system in place:

> It was clearly regulated that we have *a right* to upgrade our qualifications all five years. These workshops were held in Ulaanbaatar and we could count on being invited to attend them. Now, they say that this system is better, because they can update us annually on recent developments in education. In reality, however, this means that we have to attend short meetings with methodologists in the province and are excluded from those workshops in Ulaanbaatar that are moderated by real experts.

During a meeting with 29 teachers from the provinces conducted in April 2002, 25 reported that none of the teachers from their respective schools attended in-service training offered by the state. Only a few teachers that were on good terms with the principal or with the provincial education authorities received vouchers. However, these teachers were forced by their principal to use the vouchers for Independent Learning. In doing so, they purchased teaching material that would benefit the entire school, rather than the individual teacher.

Methodologists

The methodologists who, according to teachers, were the main beneficiaries of the voucher system, had surprisingly similar complaints. Most of them found the training, moderated by staff from the Ministry of Education or by university professors with little or no teaching experience, equally abstract. Informants said: "How are we supposed to replicate what we learned in Ulaanbaatar in our provinces? The teachers will think that we attended the workshops only because they were held in the capital." The question of who was eligible to attend workshops in Ulaanbaatar—and who was excluded from them—was a recurring theme. Due to massive urbanization, most educators in the countryside have relatives and friends residing in Ulaanbaatar whom they used to visit during the workshops. The capital has not only become a hub for social networks, but also a center for goods and services that are difficult to obtain in the provinces.

School Principals

The principals and the education authorities in the provinces, in turn, emphasized the bureaucratic inefficiencies of the voucher system. "We are supposed to receive the list of workshops that are being offered in March and submit the list of trainees in May. Year after year, however, the ECC director has received neither the vouchers nor the program until much later in the year." In their opinion, they were trying to be as indiscriminate as possible when nominating teachers to attend, but were limited by the fact that "there simply aren't enough vouchers to send every teacher to workshops in Ulaanbaatar."

Government Officials

Surprisingly, mid-career and senior staff at the Ministry of Education expressed concerns about the feasibility of the voucher system as well. "We cannot afford it," "the nominal value of each voucher is too small, and thus too unattractive for trainers," and "it requires a lot of coordination," were the most common reasons for their dissatisfaction. Annually, the Ministry of Education is supposed to receive T 160 million (approximately $142,000) from the Ministry of Finance in vouchers for in-service training. According to an official who oversees the coordination of the voucher system, either the Ministry of Education does not receive that amount from the Ministry of Finance, some of that money disappears by the time it reaches the provinces, or the principals are not using all the vouchers that they are given. Whatever the reasons, only T 100 million ($89,000) had been spent on in-service training of teachers and administrators by the end of the fiscal year.

The (non-) implementation of the voucher-based reform took an interesting turn in the election year 2004. The MPRP-led Ministry of Education secured a great deal of public support by extending the school curriculum from ten to eleven years. Although the voucher-based policy remained in effect, the actual budget was entirely absorbed by the state-run in-service training on the newly extended curriculum. Principals, assistant principals, and only a select group of teachers benefited from these state-run courses. The majority of teachers had to choose an alternative route to professional development: they paid for their own in-service training.

Beginning in 2004, the 23 public and private colleges and universities offering preservice teacher education programs discovered that teachers are willing to pay fees for in-service training provided that the courses bear credit and have, in the medium or long term, an impact on teacher salaries. Teachers flocked to courses that make them eligible to gain additional income either by teaching after-school classes and/or earning a certificate or a Masters degree through the accumulation of credits, which by default moves them up the pay scale. Most likely this creeping privatization of in-service training was neither planned nor expected. Rather, the privatization reflects the lack of state-issued vouchers on the one hand, and the need for colleges and universities to generate additional income on the other.

Semantic Changes within the Grammar of Mongolian Teacher Education

Having explained how voucher reform was manifest at the three policy levels—policy talk, enactment, and implementation—we are led to ask why the introduction of vouchers went unnoticed among Mongolian educators. Why was the original idea, so celebrated in public, diluted when it came to policy action? And why did it evaporate once it was implemented?

The simplest explanation for these kinds of discrepancies is technocratic: ambiguity of policy guidelines and poor management. The remedy would be to improve policy design, outline the implementation procedure in greater detail, and establish accountabilities. It would be naïve to assume that this remedy had not been attempted. At various stages of the implementation process, the Ministry of Education solicited suggestions from Mongolian and international education experts on how to improve the voucher system, and was by no means short on advice (e.g., Sander and Narmandakh 1998; Narmandakh 1999). However, technocratic explanations for discrepancies between the three policy levels are both paternalistic and shortsighted. We offer two alternatives: the first is embedded in an interpretive framework of "grammar of schooling" (Tyack and Cuban 1995), and the other draws on the concept of "flags of convenience" (Lynch 1998: 9; see also Steiner-Khamsi 2005b).

Despite the radical rhetoric surrounding new reform programs, policies rarely prompt immediate and fundamental change. Any reform, within a short period of time, becomes incremental and changes are made step by step, if they are made at all. This transformation has to do with the agents of reform and, more precisely, practitioners who imaginatively interpret, adapt, and selectively implement it. Thus, the question of how reforms change educational institutions is too ambitious and doomed to lead astray. Recognition of this transformation provides many lessons for policy studies, forcing us to expand our horizon in two directions: we need to adopt a long-term perspective to assess the impact of reforms, and we must abandon the belief that a paper, a decree, or a policy alone can change the status quo. There is much to gain by leaving behind a top-down ministerial perspective, narrowly focused on the question of how (their) reforms change schools or institutions. We can then see how policies are reinterpreted, or as Cuban (1998) asserts, "how schools change reforms."

A glance at the history of educational reforms reveals a preoccupation with themes that periodically reemerge in each reform wave. David Tyack and Larry Cuban (1995) identify several recurring themes in U.S. public school reform, such as performance-based salary or school-business partnership. The phenomenon of recurring themes in education is an important element in the "grammar of schooling" (Tyack and Cuban 1995: 85ff.), and it is also an opportunity for us to engage in a contextual analysis of educational import. To some extent, reform concepts need to resonate locally with existing practices or beliefs, otherwise they will be misunderstood or simply ignored.

In Mongolian teacher education, the practice of regularly updating teachers on new developments holds a prominent place in the "grammar" of in-service teacher education. This strong belief in "lifelong learning" or "uninterrupted education" (*tasraltgui bolovsrol*) of teachers has endured since the socialist past. The grammar of teacher education was preserved, but as the following demonstrates, the semantics has changed. Although the two voucher decrees did not ignite a revolution in Mongolian teacher in-service education, it would be imprudent to only address the incremental changes that the "non-reform" instituted over the past decade. When we abandon the transformation period as our timeframe for comparison, and replace it with a larger framework that includes teacher in-service education during the socialist period (in particular from 1969 to 1990), we achieve a more comprehensive analysis of what has been added to, or omitted from, voucher-based reform. This long-term perspective elucidates how fundamental the changes in the 1990s were, as compared with previous decades.

The Mongolian system of in-service teacher education, established in 1969, was similar to what existed in the Soviet Union and other socialist countries (Sachsenmeier 1978). A prominent feature of the socialist system was "lifelong learning," which included the right of each teacher and administrator to attend centrally organized teacher education sessions every five years. In Mongolia, these four-week sessions were organized by the Ministry-affiliated School of Education Development (later established as a department in the State Pedagogical University), and were mostly held in Ulaanbaatar. In-service training for preschool and primary school teachers were first held in Ulaanbaatar, and then expanded into several provinces where the State Pedagogical university had regional branches, notably in Arkhangai, and later in Dornod and Bayan-Ölgii (Sandshaasüren and Shernossek 1981). Approximately half of the training time was allocated for methodological and subject-specific topics, and the other half dedicated to the core principles of Marxist-Leninist theory, legal foundations, educational planning, and health education.

A major change in the 1990s, reinforced in the two voucher decrees, was the abolishment of teacher education sessions in Ulaanbaatar that had enabled teachers and administrators from the provinces to be in the capital periodically. In addition to these centrally planned sessions, teachers and principals had a wide array of "decentralized" programs available, including short training programs at both regional and school levels. Moreover, the long-held socialist practice of Independent Learning was maintained, but with much fewer resources. Until the mid-1980s, sufficient funds were available to equip the schools and provincial centers with libraries that were conducive to independent research and learning by teachers and principals.

In practice, the postsocialist voucher reform employed the same concepts that were utilized during socialist times. Postsocialist discourse simply inverted the socialist connotation of in-service education. These semantic changes deserve elaboration. The concept of "lifelong learning" under socialism made it mandatory for teachers to attend a relatively extensive teacher education program at least every five years. In contrast, UNESCO's Lifelong Learning agenda of "uninterrupted education" encouraged teachers and administrators to attend shorter in-service training sessions annually, most of which were only offered in the provinces. The explanation for dropping the mandatory four-week courses, and replacing them with optional, truncated two- to three-day workshops, was unconvincing to many. Policy makers claimed that Mongolia was undergoing massive social changes and that therefore it was necessary for teachers and administrators to attend annual workshops.

Double-Talk and Political Schizophrenia

Mongolian vouchers have very little in common with voucher models in other countries. Many would agree that replacing registration forms with vouchers does not typically qualify as "voucher-based reform." In Mongolia, the policy talk on "decentralization" meant, at the level of policy enactment, that the new in-service training policy also provided options to attend workshops outside the "center" (Ulaanbaatar). "Breaking the state monopoly" alluded to the regulations of 1998, wherein three types of state institutions (universities, colleges, and research organizations), rather than one (School of Education Development) were permitted to offer teacher training. "Choice" and "individual choice" were phrases newly introduced to denote that teachers had to identify their subject matter on the voucher so that they could be invited to subject-specific workshops rather than general workshops. In other words, the vouchers were treated as registration forms for tightly monitored workshops offered by three types of state institutions. Thus, the voucher reform in Mongolia is an example of discursive borrowing at the level of policy talk, with little consequence for policy implementation.

The voucher idea of the late 1990s sailed under the postsocialist flag of quality improvement, drifting away from the socialist notion of equal access. By allowing multiple providers to compete for customers, the quality of the training "product" was supposed to become superior. Additionally, the voucher-based reform abolished the socialist conception of a universal *right* and *obligation* to in-service training, replacing it with a cascade model. Methodologists were first trained as trainers by recognized experts in the capital, and then required to teach a shortened version of what they had learned to teachers in the countryside.

The term "flag of convenience" (Lynch 1998: 9) well summarizes a popular practice: Ministries in low-income countries use catch phrases that resonate with international donors in the hope of attracting international funding. Shortly after the funding has been secured, the project staff allocates the bulk of resources to other objectives. We propose the buzzwords "quality improvement" and "market orientation" for postsocialist countries to complement the list of catchphrases—multiculturalism, girls' education, community involvement—that James Lynch (1998) has enumerated for developing countries. Once "the veil of deception is cast aside and the 'deep structure' of the project is examined" (Lynch 1998: 24), it becomes apparent that the projects sails under a different flag.

In Mongolia, however, there has traditionally been a huge gap between what is done versus what is said to different audiences. The discrepancy between policy talk and policy implementation was institutionalized in socialist colloquial speech, often referred to as "political schizophrenia." Citizens became accustomed to a syndrome manifested in a mismatch between political announcements (policy talk), ministerial decrees (policy enactment), and implementation. The fact that political announcements and ministerial decrees were utterly unrelated to concrete guidelines and implementation strategies is due to a difference in target audiences. Political announcements and decrees were meant for an "outside" audience, as opposed to implementation guidelines, which were circulated among an "inside" audience. Political announcements were directed toward other "fraternal socialist countries"—political allies abroad—and provided the required lip service for proletarian internationalism. This, in the case of Mongolia, resulted in generous external assistance from the socialist camp. In contrast, guidelines and implementation plans target the local population.

It is indicative of enduring political schizophrenia that in Mongolian, the word "policy" is difficult to translate. There are at least two expressions used to express "policy": one is commonly translated as "national program" and the other as "guideline."[4] Similar to political announcements of the socialist era, today's national programs are grandiose political announcements of unrealistic government strategies with impressive target figures, intended for potential funders of reform plans. A good case in point is "The National Programme for Preschool Strengthening" adopted by the government on April 10, 1995. The Ministry of Education established a five-year plan (1995–2000) for expanding access to preschool education in Mongolia from 20 to 80 percent, and improving the overall quality of preschool. The Ministry of Education used the national program to secure international funding. Three international organizations, Save the Children UK, the Mongolian Foundation for Open Society (Soros Foundation), and World Vision contributed funds, but not nearly enough to generate the quadruple enrollment rate that the national program had promised (Batdelger, Dulamjav, Enkhtuya et al. 2000). When the program ended, the gross enrollment in preschool education was 27.3 percent—8.5 percent higher than in 1995, when the National Program for Preschool Strengthening was proclaimed—but still slightly lower (0.6 percent) than in 1991, when enrollment started to drop drastically (MOSTEC, UNDP, UNESCO et al. 2000: 38). More importantly, the increase in enrollment was seven times less than the national program had projected in 1995. In an attempt to redeem itself in the eyes of international donors, the Ministry of Education embellished the outcomes at the end of the program by indicating that an additional 14.3 percent of preschool-aged children had been enrolled in "out-of-classroom training" (MOSTEC, UNDP, UNESCO et al. 2000: 31)—short-term preparation courses offered over the summer months, which last four to eight weeks. The addition of these courses increased the total percentage of gross enrollment in preschool education to 41.6 percent, still only half of the initial target.

From the Mongolian perspective, national programs are long-term strategies directed toward the international donor community. They signal the general direction that the government is committed to undertake in the long run *if* it receives sufficient external funds. In contrast, guidelines deal with the current situation and target implementation. There is great skepticism among the population toward national programs—five-year plans, seven-year plans, and other multiyear plans—because there is little evidence that they are ever rigorously pursued, let alone fully implemented. As mentioned in chapter 5, the education sector strategies of the 1990s replaced the socialist multiyear plans of an earlier era. Concerned that the strategies would be perceived as shallow promises, the authors of the 1993–1994 Masterplan implored the reader not to mistake the long-term strategies for yet another multiyear plan (Mongolia Ministry of Science and Education and Academy for Education Development 1993–1994: 3). The first Education Sector Review (Government of Mongolia 1993: iii) reminds the reader, "[V]ague policy or planning pronouncements, and those that are unaffordable, should be avoided in favor of practical and realizable objectives."

Taken together, national programs and guidelines come across as double-talk, dishonesty, or flags of convenience. The first accusation is frequently heard among international experts working in Mongolia. In contrast, local policy makers and practitioners put national programs in perspective, and acknowledge that they are written—as were the Five-Year, Seven-Year or Ten-Year Plans under socialism—for political and economic purposes. What matters most to them is whether such national programs secure international approval and funding, and only secondarily whether and how they ever get implemented.

Shift from Institutions to Networks

National programs typically function as political "education-for" campaigns such as Preschool Education for All, Quality Improvement for All, and so on. In the case of the 1998 vouchers-based campaign, however, the signal was mixed. From the long-term perspective it must be acknowledged that "the age of the market" evokes negative connotations for those who experienced the early days of the transformation period. Buying access to resources has become a feature of many postsocialist countries, including Mongolia (Ledeneva 1998), and it is rumored that government officials, their friends, and the friends of their friends were the main profiteers of privatization. A plethora of bureaucratic hurdles such as the need for proper registration, licenses for opening private enterprises, state permission to do construction work, and so on, have been put in place to limit access to resources, tightly controlled by government officials. The right connections, networks, or cash are means to circumvent these limitations.

The "shock therapy" of the early 1990s—decentralization of planning, abolition of price control, and privatization of state property—made a deep impression on those who did not know the right people at the right time. It is not surprising that in Mongolian the term "market economy" (*zakh zeel*) provokes associations with "first-come-first-serve," where a few benefit, and the majority are left behind. Unequal access is a feature of many new market economies. Caroline Humphrey and Ruth Mandel made similar observations in several regions of the former Soviet Union (Humphrey and Mandel 2002: 6), where privatization (*privatizatsia*) is cynically referred to as *prikhvatizatsia* (from the Russian word *prikhvatit'*, translated as "to grab").

As mentioned before, the teachers in our interviews conveyed that they had no access to vouchers because principals distribute them in an unfair and nepotistic manner. Of special interest for this study are anthropological examinations of popular reactions to the emergence of bribery, fraud, and corruption in the absence of clear legislation and institutional accountability (Verdery 2003; Humphrey 2002b; Sneath 2002b). In Mongolia, for example, it is now more common to resort to social networks than state institutions for assistance. The latter have lost authority for a variety of reasons. In election years, more than 9 percent of all civil servants, including principals and education authorities, are replaced along party lines (World Bank 2002: iv). The high turnover rate in state institutions reveals a long-standing system of political patronage, and contributes to the public's general assessment that state officials are neither trustworthy nor accountable for their actions.[5]

In his analysis of social reciprocity and obligation in Mongolia, David Sneath (2002b) conducted two studies dealing with "gifting" (giving gifts), bribes, and corruption. These studies draw from a government-sponsored survey on corruption, which includes responses from 1,500 Mongolians, and his own explorative study based on interviews with 140 residents of Ulaanbaatar and a rural district in Arkhangai (Sneath 2002b).[6] The majority of the respondents from the survey (70.2 percent) found corruption widespread in the 1990s, and only very few (7.2 percent) considered it a feature of the previous socialist era. The two most frequent reasons given for widespread corruption during the 1990s were "officials are not sufficiently bound by their duties," and "dishonest privatization of property" (Sneath 2002b: 86). It is a common sentiment in Mongolia that, to get things done, one needs to engage in all kinds of gifting practices, ranging from honorific expressions of gratitude to illegitimate practices of bribery. The informants in Sneath's study listed the following six examples as occasions for gifting *(shovgor*[7]*)*: to secure a place for their daughter or son at a university, to receive the necessary grade on an exam, to get a job, to complete official documents or papers, to get a bank loan, and to facilitate a business transaction (Sneath 2002b).

Sneath's ethnographic account of monetized social interactions in Mongolia is precise, and it includes a description of the areas Mongolians consider nebulous. What is considered an acceptable expression of gratitude and gift giving, versus what qualifies morally as an unacceptable practice of reciprocity and bribery, are matters of great complexity and subject to social change. To illustrate his point, Sneath comments on a newspaper article with the headline "Give Doctors and Teachers Bribes! Why Not?" The author of the article makes a case for differentiating between the recipients of a bribe. If the recipients are respectable individuals to whom one is socially indebted, such as doctors who save lives or teachers who "taught us our professions," then a bribe is morally justified. Teachers and doctors are not only members of traditionally respected occupations, but also earn low salaries: the bribe is justified.[8]

In our own study in the eastern province of Dornod (Steiner-Khamsi, Stolpe and Tümendelger, 2003), we made similar observations. We found that the provincial education authorities were very sympathetic toward teachers who earn additional income by teaching extra classes, tutoring students individually, and receiving gifts from parents. The provincial education authorities neither sanctioned these practices nor tacitly tolerated them in the form of a "suspended punishment" (Ledeneva 1998: 7; see also Rasanayagam 2003); they were, however, eager to regulate them. Such regulations specified which teachers were allowed to earn an additional income, thereby ensuring that parents only invested in those teachers that the education authorities deemed qualified.

As discussed in the earlier section of this chapter, teachers have misgivings about the voucher distribution practices of principals. In an environment where the right of every teacher to in-service training is made available to only a select few, it naturally becomes an object of great speculation as to which teachers have been selected. Given scarce resources and in the absence of clear selection criteria, most principals in this study resorted to their own social networks, and registered those teachers who were either relatives or friends, or were senior teachers who could reciprocate the gift of being sent to Ulaanbaatar.

In Search of Voucher Reforms that Work

Despite sailing under the postsocialist flag of quality improvement, the 1998 voucher reform failed. The government did not secure international funding for the professional development of teachers, and the voucher program had the same destiny as countless other unfunded national programs that were eventually discarded. The flag of convenience was hoisted at half-mast: the language of the reform was preserved, but no further attempts were made to actually implement it.

Unlike the structural adjustment reforms presented in chapter 5, notably decentralization of educational finance and governance, the voucher reform was not externally imposed. However it does constitute a fascinating case of a voluntary reform import. It was a case in which international organizations did not exert direct pressure on the Ministry of Education of Mongolia to reerect the dilapidated in-service training system in line with experiences from elsewhere. Instead, they directly imported in-service training programs for teachers, using their own international trainers, materials, and funds. The Ministry of Education, in turn, channeled its loan from the Asian Development Bank (ADB) into the training of administrators, school principals, and assistant principals in the capital city of Ulaanbaatar (Spaulding, Boldsukh, Munkjargal et al. 1999). This targeting of administrators left the field of teacher in-service training wide open for regional education authorities in the cities and the provinces, and for international nongovernmental organizations (NGOs). There are three prominent international NGOs involved in teacher in-service training: the Danish International Development Assistance (DANIDA), the Soros Foundation/Mongolian Foundation for Open Society (MFOS), and the Save the Children Fund UK. NGOs established a division of labor among themselves that targeted different groups of educators. Accordingly, DANIDA focused on primary school teachers, science teachers, and math teachers, the Save the Children Fund UK targeted preschool teachers, and MFOS focused on teachers in the humanities and social sciences. The donor-run in-service training system for teachers and the state-run system for school administrators operated in tandem. The international NGOs hoped that the state-run system would eventually incorporate a few of the prominent features of the donor-run system such as school-based training, practice orientation, and interactive training methods.

Toward the end of the 1990s, these international NGOs redoubled their reform efforts and actively sought ways to sustain their previous efforts in in-service teacher training. It was only in 2000 that MFOS, one of the largest donors in education in the late 1990s, discovered that a voucher-based system had already been in place for the past two years. At that time the majority of Mongolian experts in education knew of the Ministry's intention to introduce voucher-based reform. But only a few were certain that it had indeed been implemented. For MFOS and other international NGOs, it became a high priority to confirm whether they could integrate their in-service training programs in the voucher-based system, certify their trainers with the Ministry of Education, and pay them through vouchers.

It was against this background, and in the context of the School 2001 project of MFOS in particular, that we conducted this study on vouchers. In 2001, the trainers of the School 2001 project had been certified, but without any financial consequences for the Ministry of Education. Since 2004 teachers have enrolled in professional development courses offered by trainers of School 2001 or other NGOs. They have paid for these courses, and in return have received university credits. Additionally, the third loan of ADB earmarked funds for the current curriculum reform (extension from 10 to 11 years of schooling), and set aside money for the in-service training of primary school teachers who are directly affected by the reform. At the same time, NGOs are still in search of voucher-based models in the postsocialist region that actually work. After all, their efforts in in-service teacher training can only be sustained if the government is prepared to distribute vouchers to a free market of training providers.

The verdict among the NGO-funded training providers is that the voucher model of Samara, Russian Federation, was actually successful. The Soros Foundation Kyrgyzstan initiated a project on the "voucher system in in-service teacher training," which matched funding from USAID.[9] The funding is used to pilot a voucher model in the Kyrgyz region Issyk-Kul Oblast tailored after the experience in Samara, and will introduce a new financing mechanism for in-service teacher training by 2007 (Soros Foundation Kyrgyzstan 2004). The Samara model claims to facilitate an individualized education path (IEP) which enables teachers to accumulate courses toward earning a higher rank or degree. As with the 1998 Mongolian voucher reform, the voucher system in Samara encourages teachers to create their IEP with "the freedom to select content, form and the provider" (Pacurari, Batkhuyag, and Mason 2004: 25). Samara is likely to become a place of pilgrimage for NGOs seeking proof that vouchers can actually be put to use for teacher in-service training. Within the

OSI network, a team of experts from the Kyrgyz Soros Foundation made a study visit, determined to make the positive experiences from Samara known to policy makers in Kyrgyzstan. These Kyrgyz experts report that the Samara voucher system boosted the market for in-service training, and that more than 15,000 teachers have attended voucher-based training since 1997, when the voucher system was introduced. Teachers receive vouchers every 5 years for an in-service training program that lasts 144 hours (approximately 4 weeks). Half of the training program consists of "fixed units" in which the trainees are taught in the "competence-based method of education" and information technology, and the other half consists of "variable units" determined by the teachers (Department of Science and Education of the Samara Region 2001: appendix 1).

Utilizing a reform from the Russian Federation as a model for emulation and as a means to generate reform pressure domestically seems to work well in Kyrgyzstan. Similarly, the Mongolian Foundation for Open Society also directed its attention to other postsocialist countries in Central Europe and the Baltic states (especially Lithuania) to "fix" voucher reforms that went wrong. This East-East orientation has not been the case for government officials in Mongolia. To date, they have been finance-driven, and their points of reference for teacher education reform have been exclusively donor countries such as, Australia, United States, United Kingdom, Italy, Netherlands, Portugal, and Scotland (Narmandakh 1999).

An interesting turn has occurred with NGOs in Mongolia entering the scene of state-funded in-service teacher training. Rather than preserving their own, expensive parallel system, they aim at mainstreaming NGO-run professional development into the state-funded system. This handover or transfer from NGOs to government structures is the rule, and not the exception. Most international NGOs in Mongolia assign themselves the role of incubators in that they pilot, develop, and fund a new practice, which the government eventually is supposed to borrow and implement nationwide. In fact, sustainability and effectiveness in the world of international NGOs is often defined in terms of such a transfer. An NGO project is considered successful when the government has adopted it, and continues to finance it either with state funds or with financial means from external assistance, notably with loans from the Asian Development Bank or the World Bank. An analysis of the interaction between international NGOs and the government helps tease out the complexity of policy import. For the past seven years, the government did not move beyond discursive positioning. It placed itself in an imagined international space in which presumably all other educational systems have implemented market-oriented reforms identical or similar to the voucher-based reform in Mongolia. As beneficiaries of a well-functioning system that would liberalize the provision of teacher training by means of vouchers or another reform, the NGOs in Mongolia are put in a difficult predicament. Should they support or insist upon the realization of voucher-based reform, or should they accept that this reform is beyond repair and was never meant to move beyond lip service?

9

WHAT IF THERE IS NOTHING TO BORROW? THE LONG DECADE OF NEGLECT IN NOMADIC EDUCATION

Unlike the case studies of imposed or voluntary borrowing discussed in the four preceding chapters, this chapter deals with the absence of borrowing in the area of nomadic education. An exploration of this transfer vacuum helps us to understand the logic of international donors who, in a long decade of neglect (1991–2003), refused to support the rehabilitation of boarding schools in Mongolia. Until 1990, a nationwide system of boarding schools made it possible for children of nomadic herder families who lived in remote rural areas to attend school. The enrollment figures speak for themselves: By the end of the 1960s, universal basic education had been achieved; in 1990, the gross enrollment ratio for 8–15-year-olds was 103 percent, and the adult literacy rate was 97 percent (Government of Mongolia and UNDP 2000: 26). By all accounts, the Mongolian boarding school system was a "good," or even "best practice" that, during the country's decade of transformation in the 1990s, went astray. Taking a cue from scholars who suggested that we turn our attention to practices in development that went right (see Vavrus 2003), we briefly sketch the features of the socialist boarding school system, and analyze its transformation during the postsocialist period.

NOMADIC EDUCATION IN SOCIALIST TIMES

In Europe and North America, boarding schools tend to evoke horrific associations of cultural alienation and forced assimilation targeting disenfranchised minorities, notably Roma and Native Americans (Connell-Szasz 1979; Child 1998). This is not the case in Mongolia. During the socialist period, several features of Mongolian boarding schools made them popular among nomadic herder families. They were: (1) a continuation of a previously existing organizational structure of schooling, (2) child-friendly, (3) integrative, and (4) close to the families.

First, it is important to bear in mind that, ever since the country's successful conversion to Buddhism in the seventeenth century, monastic schools in Mongolia were the first institutions that enabled access to education for a population that, for a long time, was almost entirely composed of nomadic pastoralists. It was customary for herder families to place at least one of their sons, between the ages of 7 and 10, in a monastery, where they would remain for a few years to learn Mongolian and Tibetan script, Buddhist science, arts, and handicrafts (Shagdar 2000; see also chapter 2). Thus, in the 1930s, when the first state-run boarding schools were established in postrevolutionary Mongolia, the organizational structure of boarding students in the location of the school was neither new, nor alienating.

Second, unlike monastic schools, the state-run boarding schools attempted to create an atmosphere that was child-friendly and family-oriented by allowing siblings, relatives, or neighbors (boys and girls) to reside in the same room. Two additional factors, in particular, ensured that the children did not emotionally distance themselves from nomadic pastoralism and continued identifying with the parents' lifestyle: The relatively late school-entrance age (eight years old) and the flexible schedule for school vacation (the latter was later on abolished), seasonally adjusted to the work schedule of herders. As a result, the children had ample opportunity to experience the lifestyle of nomadic pastoralists.

Third, the boarding schools were not separate institutions, but an integral part of the socialist education system. Located in the administrative centers at the province (*aimag*) and district levels (*sum*), they accepted all children from the area that needed accommodation at the school: children of workers, civil servants, employees, and members of the agricultural farms (*sangiin aj akhui*) or animal husbandry collectives (*negdel*).

Finally, the first legislation in Mongolia concerning compulsory education dates back to the 1920s, but it took another 30 years to actually implement it. The system lacked not only teachers, but also the funds to build schools and boarding schools. From the outset, it was uncontested that boarding schools were a precondition for reaching out to children of nomadic herder families, and other families living in remote rural areas of the Mongolian steppe and desert. What exactly triggered the turning point in the 1950s, when schools and boarding schools mushroomed in all the provinces of the country? The most influential domestic factor was the collectivization of livestock and the establishment of agricultural farms in the 1950s. As state institutions, the animal husbandry collectives and the agricultural farms overlapped with the administrative units and thereby functioned as an organizational structure, overseeing both the economy and the infrastructure in rural areas, including schools. As a result, many boarding schools were established with financial support from the collectives (Schöne 1973: 81, 158). The *negdels*, in particular, understood quite well that animal husbandry in Mongolia depended on the ability of herder families to preserve their nomadic lifestyle. From the early 1960s to the early 1970s, the collectives covered, on average, 10 percent of the total cost for erecting schools and boarding schools in rural areas (Ardyn Bolovsrolyn Yaam 1976: 75, 183–193).

The enormous cost of operating and maintaining the boarding school system has been a contentious issue since the first days of compulsory education in Mongolia. After 1990, the Ministry of Education lacked not only the funds to fix the dormitory buildings and replace the furniture, but also to cover the maintenance fees, notably the high costs of electricity and heating. Starting in 1991, boarding schools became notorious health hazards for children; they were seriously under-heated and in poor hygienic condition. As a result, many boarding schools were closed down completely, and most boarding schools that remained in operation were in a deplorable state: under-heated, with leaking roofs, broken windows, and no electricity (see also Steiner-Khamsi and Stolpe 2005).

THE TRAGEDY OF BOARDING SCHOOLS

In 2003, when we visited settlements, schools, and boarding schools in the eastern province of Dornod, we saw a bleak picture: The boarding schools in Dornod had ceased to be integrative, child-friendly, or geographically close to the family. The few boarding schools that remained in operation throughout the 1990s were in poor physical condition, and the few students that were admitted during that period, of the many who applied, did not have fond memories; thus, parents were seeking alternative solutions for their children.

Using the province of Dornod as a case study, we investigated the living arrangements that parents chose for their children during the school year in great detail. For each of the 14 district schools in Dornod, we gathered statistical information regarding school enrollment, and if applicable, the number of students that applied to boarding schools as opposed to those accepted, the number of dropouts, information on school finances, and the ranking of schools with regard to their placement in *olympiads* (competitions) and standardized eighth-grade and tenth-grade examinations (see Steiner-Khamsi, and Tümendelger, Stolpe 2003). We also used the qualitative data we had collected from interviews with teachers and principals from 19 schools (5 province-center schools, 13 rural district-center schools, 1 village school), and from visits with 34 families from 12 different rural districts (*sum*). Additionally, we interviewed the staff at the Education and Culture Center in the capital of the province, that is, the education authorities at the provincial level, and met with several governors of rural districts and villages.

In our study on school-related migration in Dornod, we paid special attention to "school-year migration"—when school-aged children stay with one of their parents, another relative, or with close friends near a school. In our selection of interview excerpts, we narrowed our focus to the type of living arrangements that parents secure for their children, depending on the kind and size of (financial and social) resources available to them. The first excerpt is an example from a resource-poor (*bolomjgüi*) family, whereas the second presents a living arrangement from a family with resources (*bolomjtoi*).

Example from a resource-poor nomadic herder family:

The education of our children is very important for us. We would even be prepared to move to the *sum*-center [school in the rural district-center] over the school year, if we could. But I need to help my husband to look after the 200 livestock. This year, I also got an additional job here, for which I earn T 32,000 [$28] per month. Every morning, I have to measure the temperature of the lake, and send the data for the weather forecast to an office in Choibalsan [province-center of Dornod]. I would have preferred to place my children in the boarding school in the *sum*-center rather than leaving them with friends. Unfortunately, the condition of the *sum*-school and the boarding school is very modest (*taaruukhan*). The school places the boys in one classroom of the school building, and the girls in another classroom in the kindergarten building. The school hasn't been renovated for years, the roof is leaking, and most classrooms can't be used over the winter, because the heating is broken. Moreover, they don't have water for the children to wash themselves, and many of them got fleas. Therefore, we decided to place our oldest son with our friends in the *sum*-center, but after the first grade he was so homesick that we had to take him out of school. For the next two years he stayed at home and helped with the livestock. Now that his sister has entered school too, we re-enrolled him. They are now both in the same third grade class, and are able to look after each other. They are both happy about this solution even though they are two years apart age-wise.

Example from a nomadic herder family with resources:

We are very busy throughout the year. With 1,000 livestock heads (all 5 species—sheep, cows, camels, goats, and horses), we have several of our children helping us. Two of our sons need to move up to ten times a year to do *otor*.[1] We are very pleased that we also have children living in Choibalsan and Ulaanbaatar. This way we know what is going on. We also subscribed to newspapers, have TV, and a generator for electricity in our *ail* [temporary settlement]. Small children need their mother. That's why two of our youngest grandchildren stay with their mother in the *sum*-center during the school year. Their father (our son) helps us with the livestock throughout the year. In the summer, my son, his wife, and their two children are re-united, and live together with us in the countryside.

The principals, teachers, and dormitory staff we interviewed all pointed to the following, apparently unprecedented phenomenon: boarding schools are now predominantly used by families who are resource-poor, that is, have neither sufficient income and possessions to purchase or rent a second residence nor access to individuals with resources (relatives, friends) living near a school. In all but one school, the number of applicants for a space in boarding school exceeded the number of spaces available. Realizing that families without resources cannot afford to make living arrangements for their children on their own, many schools give priority to applicants from poor, nomadic herder families. Several schools even accommodate the children of poor parents who live in the district-center, but do not have the means to provide a healthy environment (heating and meals) for their children.

What we found particularly puzzling was the extreme social stratification evident in living arrangements for school-aged children: Families who were either poor, or lacked social networks in the district- and province-centers where the nearest school was located (so-called resource-poor families), tended to place their children in boarding schools. As a result, the boarding schools that we observed predominately accommodated students from resource-poor families who had little income or few possessions, or who could not draw from a network of kin or friends in a financial position to raise additional children in their households. At the other extreme were families with resources who had been able to maintain their primary residence in a remote rural area, and purchased an additional apartment near a grade 1–10 school, located either in the capital of the province or in the nearest district-center. Between these two extremes—placing children in a boarding school or placing them in a second, privately owned residence—existed a wide spectrum of living arrangements parents could choose from, based on the age of the child, the number of younger and older siblings in a family, the distance to the school, and several other factors (see Steiner-Khamsi, Stolpe, and Tümendelger, 2003).

Many reasons account for the transformation of the boarding school from a universal institution serving all students, to its postsocialist incarnation as a particularistic system, predominately accommodating students without better options available. As mentioned before, the government lacked external funding to cover the maintenance costs for running the boarding schools, let alone to invest in the renovation of buildings from the 1960s and 1970s in urgent need of repair. The government response to this financial impasse was twofold: In the first half of the 1990s it either completely closed down, or reduced use of boarding schools in need of major repair; in the second half of the decade (1996–2000) it charged the meals of boarders to their parents—a policy better known as the "meat requirement." The Meat Requirement prescribed that a family pays for 154 pounds of meat per child a year (equivalent to 2 or 3 sheep), an amount that low-income herder families could not afford and which, as a consequence, forced many poor children out of school. Other researchers identified a dual cause for the high dropout rates during the 1990s: High-income herders withdrew their children (mainly sons) so that they could help with animal husbandry. Low-income herders, in turn, withdrew their children from school because they could not afford the private cost of schooling, and in particular, were not able to satisfy the Meat Requirement (Stolpe 2001). Not surprisingly, the dropout rates for children of poor herders were highest from 1996 to 2000, during the period of the Meat Requirement policy.

The following table 9.1 illustrates the dramatic decrease in boarding school enrollments throughout the 1990s, and a gradual recovery after 2000, when the Meat Requirement was abolished.

Table 9.1 Boarding School Enrollment, Period 1990–2002—Figures for Mongolia

Year	1990	...	1994	1995	1996	1997	1998	1999	2000	2001	2002
Students	64,362	...	21,364	18,367	17,649	18,369	18,832	19,567	27,435	27,978	33,649

Source: MOSTEC 2003.

To some extent, these figures also reflect the demographic changes that occurred during the 1990s, in particular, the emigration from rural and semi-urban areas (where boarding schools are based) to the capital of the country (Ulaanbaatar) and other urban areas (National Statistical Office of Mongolia 2001).[2] Unequal access to resources (including electricity, employment, health and education services), unequal communication and transportation networks, and, in general, the unequal living standards between rural and urban areas, all account for the huge internal migration in Mongolia during the past decade. The rural flight was exacerbated by a series of natural disasters between 1993 and 2004 that deprived many herder families of a source of income.[3] In short, the villages have become more rural and remote than they were during the socialist era.

As mentioned before, Caroline Humphrey and David Sneath (1999: 179ff.) distinguish between "urbanism" and "urbanization" to characterize developments in postsocialist Inner Asia. Whereas urbanization reflects the strong concentration of a population in a particular location, urbanism indicates easy access to both social infrastructure and essential goods. According to many accounts, during the socialist period the majority of Mongolian villages had access to electricity and safe water, and had an infrastructure—a school, a post office, a hospital, a veterinary post, a library, shops, and a cultural center—similar to any city district. Additionally, villages could rely on a well-functioning air transportation system to connect them to the provincial center and the capital (Bruun, Ronnas, and Narangoa 1999). These features of urbanism existed wherever there were residents working in animal husbandry collectives (*negdel*), or in state agricultural farms (*sangiin aj akhui*). Along with the decollectivization of *negdels* and *sangiin aj akhui*, the organizational structure for providing an income and maintaining an infrastructure in rural areas gradually dissolved.

In hindsight, it is difficult to distinguish between cause and effect: Were rural schools, and boarding schools in particular, "left behind" as a result of the rural-urban migration, or did the abandonment of rural development trigger the emigration from remote rural areas? In fact, it seems that reasons for abandoning the villages are intertwined, creating a situation that gets progressively worse. During the early days of the transformation, residents flocked into the cities and semi-urban settlements in search of employment and better living standards. After a critical mass of residents, including "good" teachers, had left the villages, school-related migration became an issue. In the past few years, rural schools have struggled to retain both teachers and students, as school-related migration (Steiner-Khamsi, Stolpe, and Tümendelger, 2003) has become a decisive factor for deserting rural areas. A study noted that among the new immigrants in Ulaanbaatar and three other urban locations, the prospects of "studying," or creating a better future for one's children, scored among the top three factors (the most important was "work") that propelled them to move (Bolormaa 2001).

During the socialist period the boarding school system in Mongolia functioned well, because it was free, integrative, and universal. Once a large number of boarding schools were shut down, either partially or completely, and the staff dramatically reduced, the education system faced a tragic situation. With fewer boarding schools the pedagogical conditions grew worse, and there was greater pressure to select only those students who were desperately in need of accommodation. As a result, the boarding schools have ceased to be (geographically) close to the majority of herder families. Instead they have become a substitute for *bolomjgüi* families who had no financial means or social networks to accommodate their school-aged children.

Best Practices in Nomadic Education

The previous section dealt with the tragedy of the boarding schools during the postsocialist period. Specifically, it explored their transformation from universal institutions into institutions that, due to lack of funding and space, are only able to accommodate the neediest of the needy. Could the tragic turn of the Mongolian boarding school system have been averted? More specifically (1) were there alternative forms of schooling for children of nomadic families that would have been worth pursuing, and (2) was the boarding school system of the socialist period really a "good" or "best practice" that would have been worth preserving in some way?

It might be worthwhile to temporarily suspend our small unit of analysis, and introduce an international, comparative dimension to the discussion. After all, the Mongolian education system is hardly unique in terms of having to ensure access to education for a population that is both dispersed and mobile. Arid land that requires herders to move their livestock periodically to another pasture is more widespread than commonly assumed. The academic literature on nomadic pastoralists is extensive; a systematic and comprehensive account of nomadic pastoralism has been written by Anatoly Khazanov (1994), and a comparative study on nomadism in Inner Asian countries of the 1990s has been compiled by Caroline Humphrey and David Sneath (1999). Unfortunately, compared with ecological, geographical, and social anthropological studies (see Fratkin 1997), educational research on nomadic pastoralism is rather sparse.

Caroline Dyer (2001) discusses nomadic education against the backdrop of the 1990 international UN declaration "Education for All," and criticizes its exclusive focus on schooling for sedentary populations or rather, its disregard for the formal education needs of nomadic groups. There exists only one comparative literature review that explicitly deals with nomadic education in different societies, including Mongolia. Saverio Kratli (2000) prepared a review of the literature on the topic (with a special focus on Iran, Sudan, Kenya, Somalia, Nigeria, and Mongolia), examining the Mongolian context in great detail.

Kratli (2000) examined several educational provisions for children of nomadic pastoralists, and found the model of the Mongolian boarding schools convincing to the degree that he proposed emulating and adopting it in other nomadic societies. His investigation of factors that most likely accounted for the high enrollment rates in Mongolia during the socialist period revealed two types of factors, which we label "hard" and "soft." The hard factors that positively impacted universal enrollment include the strict enforcement of compulsory education, free access to education, the abolition of child labor, and well-equipped boarding schools. The soft factors, in turn, concern the "non-antagonistic culture towards nomadism" in Mongolia (Kratli 2000: 48), the late school entrance age (eight years), and long summer vacations. The soft factors prevented children from feeling alienated from their families and from nomadic pastoralism, despite long periods of separation from their parents. According to Kratli (2000), four characteristics of the Mongolian model were conducive to the near universal enrollment of children of nomadic herder families: (1) a nonantagonistic treatment of both nomadic and sedentary children leading to a supportive learning environment, (2) a clear legal framework for nomadic education and a strict enforcement of the laws, (3) free education, and (4) a combination of nomadic education with broader development policies for nomadic pastoralists.

Overall, Kratli (2000: 61f.) is impressed that in Mongolia, the so-called universal values, which upon closer examination are firmly rooted in conceptions of a sedentary lifestyle, have been reconciled with nomadic livelihood strategies. Acknowledging the contextual peculiarities of nomadic education in different parts of the world, Kratli nevertheless advocates for the Mongolian model from the socialist past as a best practice from which other educational systems, struggling with the difficult task of ensuring universal access to a population that is both dispersed and nomadic, could learn.

Migration and Access to Education

In this chapter we focus mostly on the *seasonal migration* of herder families that requires them to find accommodation for their children during the school year. In the past, boarding schools catered to these children; in the present, families are forced to seek, as presented earlier, solutions that depend on their access to financial resources and social networks.

Background information on nomadic pastoralism in Mongolia might be helpful to understand why suggestions for alternatives to boarding schools—mobile teachers who regularly visit students in their settlements (Government of Mongolia 1993: v), home (i.e., *ger*) schooling, multigrade primary schools, or nonformal education as a substitute for schools—have been dismissed as either "premodern," or unfeasible. At the end of 2003, Mongolia was home to 25.3 million livestock, including 10.7 million sheep, 10.6 million goats, 2 million horses, 1.8 million cattle, and 255,600 camels (Grayson and Munkhsoyol 2004: 19).[4] The various migration patterns of nomadic pastoralists reflect a complex management of suitable pastures, which are determined by the vegetation zone of the location (steppe, desert, mountain, forest), the kind of livestock,[5] and the size of the herds (Bazargür, Chinbat, and Shiirev-Ad"yaa 1989, 1992; Songino 1991; Bataa 1998; Tömörjav 1999; Tömörjav and Erdenetsogt 1999).

The most recent national study on nomadic pastoralism in Mongolia was completed in 2004 (Grayson and Munkhsoyol 2004), and covered 11 provinces and Ulaanbaatar (21st city-district and Songinokhairkhan district), and 34 rural districts (*sum*). It draws on a representative sample of 773 herder households, approximately half of which are relatively poor and possess less than 50 heads of livestock, 26.8 percent own 51–100 heads, and the remaining quarter consists of herders who are well-off with a livestock herd of over 100 heads.[6] According to the study (Grayson and Munkhsoyol 2004: 55), over 70 percent of the herder households from the eastern region, and almost 55 percent from the western region move more than 3 times a year. In all other regions, the majority of herder households only need to resettle 1–3 times a year. The distance between the old and new settlement varies considerably depending on the kind and size of livestock, vegetation zone, and more recently, land rights and trespassing problems as well (Potkanski 1993; Szynkiewicz 1993; Fernández-Giménez 1997, 2001; Janzen and Bazargur 1999; Finke 2000).

Regardless of the exact distance between old and new settlements, the distance is too far for children to commute (by foot, horse, or camel) from their *ger* to the same school throughout the school year. Furthermore, the pastures are rarely in the proximity of a rural district-center where schools are located, precluding the option of being enrolled in different schools over the course of a school year. Mobility and distance, combined with a lack of dormitories and the private cost of education, are the main factors explaining why many poor herder families (half of the herder family population) keep their children at home or take them out of school. The use of (male) child labor for herding is only applicable for families with a large herd of livestock, and is not applicable for the majority of herders who cannot afford to send their children to school. Similarly, "doesn't like to learn" as an explanation for dropout (see Egelund 2002), uncritically claimed by official statistics, masks school-induced dropouts and at the same time reflects the difficulties that Mongolian teachers face while teaching in a heterogeneous classroom that includes slow learners or children with disabilities. The farewell note of a mother captures how learning disabilities, child labor, and geographical distance are interwoven in the decision to take a child out of school:

> Dear Teacher. My son is slow in learning and thus it is impossible to force him to learn. So, I'm taking him away with me. We live in a rural area and look after each other's livestock for a living. It's difficult for me to bring my son to school all the time. Bye. (cited in del Rosario, Battsetseg, Bayartsetseg et al. 2005: 55)

The Mongolian Dropout Study was based on 538 surveys and 26 interviews in 4 provinces and Ulaanbaatar (del Rosario, Battsetseg, Bayartsetseg et al. 2005). What it found was the opposite of school fatigue or unwillingness to learn. Most dropouts in rural areas were ashamed of being out of school and pledged to return to school.[7] What they were missing in particular was the opportunity to communicate with others who are outside of the narrow familial boundaries of the herder household. Herder families do not simply consider schools as a terrain where literacy or numeracy are taught, but rather view them as sites of broader social interaction, which they are not able to provide in their secluded settlements. In our own encounters with dropouts in the provinces Ömnögöv' (Stolpe 2001) and Dornod (Steiner-Khamsi, Stolpe, and Tümendelger, 2003), we found it striking how timid, insecure, and helpless children become if they have been confined to communication only in their extended family.

The Mongolian education sector learned to cope with the challenge of securing access for a population that was mostly nomadic and dispersed. In contrast, *permanent migration* from rural to urban areas is a new phenomenon. The magnitude of migration in Mongolia was first noticed in the 2000 census report (Mongolia National Statistical Office 2001). The report lists urbanization and internal migration, along with a considerable slowdown of the annual population growth, as the most alerting demographic changes that have occurred since the last census in 1989. The publication of the census report drew considerable public attention, and was widely discussed in the media. There appears to be agreement among demographers that the following three concurrent migration flows had emerged in the 1990s:

- Migration from rural areas (*sums*) to semi-urban areas (*aimag-centers*);
- Migration from peripheral provinces to the central region of Mongolia (especially Töv province); and
- Migration to Ulaanbaatar.

More than one-third (35.4 percent) of the population registered in the capital (Ulaanbaatar) was born in other regions. Based on this figure and numerous other findings from the census data, the Mongolia National Statistical Office assumes a stepwise approach to internal migration. Typically, residents of rural areas first migrate to semi-urban areas, then to the central region, and finally to Ulaanbaatar. For those that migrated to the central region that surrounds the capital, "Ulaanbaatar was the end of the road" (Mongolia National Statistical Office 2001: 58).

The impact of internal migration and urbanization is painfully felt—but not sufficiently measured and analyzed—in the education sector: On one hand, schools in Ulaanbaatar and in the province-centers are overcrowded, sometimes with 40–50 students per class, and 3 shifts per school. On the other hand, schools in *bag-* and *sum-*centers struggle with diminishing numbers of students, and a loss of professional teachers who prefer to work in the province-center or city schools. Both phenomena, overcrowding of schools in urban settlements and underutilization of schools in rural areas, have occupied a prominent position in political debates and the media. For politicians and journalists, most of whom live in Ulaanbaatar, the concern was mainly how to protect the metropolis from masses of new immigrants from the countryside. Bureaucratic hurdles were put in place to make it difficult and costly to register in Ulaanbaatar. Until 2003, illegal residents in Ulaanbaatar faced marginalization and discrimination, and were denied access to public services including health and education (see National Board for Children, Save the Children UK, and UNICEF 2003). The question soon became whether a revitalization of rural infrastructures would effectively curtail internal migration, and help contain urban poverty that had been on the rise since the mid-1990s.

Ownership and International Assistance: Rhetoric and Reality

It is not that the Mongolian Ministry of Education did not try to draw international attention to rural school development and, in particular, to the dilapidating boarding school system. The Mongolian People's Revolutionary Party (MPRP) came to power after the first democratic election, from 1992 to 1996, and again from 2000 onward. The MPRP, with a vast constituency in rural and semi-urban areas and a strong following among educators, periodically appealed to the international donor community to allocate funding for the rehabilitation of boarding schools. In contrast, the coalition parties of the Democratic Union generally advocate less for state intervention and more for supply/demand-driven policies. Consequentially they have assumed a more fatalistic stand on urbanization and rural flight. For them, schools should follow families, and not the other way around. Rural-urban migration is a fait accompli, and all the government must do is build schools in urban and semi-urban areas to ease the tension over overcrowded schools.

The MPRP has pursued a more proactive approach to leveling urban-rural standards of living. In fact, the MPRP-led Ministry of Education has regarded rural school development as an effective strategy for combating poverty in rural areas. The education component of the first Poverty Alleviation Program (1994–2000) aimed at reducing school dropouts and non-enrollment rates by renovating rural schools and boarding schools. Although many rural schools or boarding schools (266 in total) benefited from the program, the amount granted to each school was small overall, averaging $3,000 (Government of Mongolia, World Bank, and UNDP 1999: section 5.2). The schools used these grants either to fix the heating system, leaking roofs, and broken windows, or to undertake other urgent, minor repairs that were required to stay open. However, the amount was not sufficient to invest in the structural improvement of the facilities, and to make rural school facilities attractive places for learning or living.

The 2000 MPRP Action Program that brought the party back into power prioritized rural development, and explicitly mentioned the rehabilitation of boarding schools. One of the first decisions of the MPRP-led Ministry of Education was to abolish the Meat Requirement, which required parents to supply the meals of the boarding school. The Interim Poverty Reduction Strategy Paper, developed in 2001, vows to draw much more attention to the problem of access to education in rural areas. Reportedly, in 203 out of the 307 rural district schools (*sum*), the gross enrollment was lower than

80 percent. The urgent appeal for immediate action is best illustrated in the following paragraph:

> There is a real difficulty in providing herdsmen's children with dormitory facilities. Due to low population density in the rural areas compared to the national average, there is only one school in each *sum* (rural district) and it is located 10–300 km away from the herdsmen's home. In order to increase the school attendance, dormitories (in most *sums* temporary buildings) were built at *sum* schools and the Government provided all tenants with meals and other necessary goods from 1960–1990. These measures played an important role in educating herdsmen's children. As a result, by 1989 the number of tenants of the school dormitories reached up to 75 thousand and attendance in basic education increased up to 90 percent. But since 1991 over 50 dormitories were closed due to economic difficulties related to maintaining the infrastructure. (Government of Mongolia 2001a: 18)

In the same year, at a meeting of international donors held in Paris, May 15–16, 2001, the MPRP-led Ministry of Education echoed its commitment to rural school development, and requested that more than one-third of the international educational grants ($20 million from a total of $56 million) be allocated to rehabilitate the buildings of secondary schools and boarding schools (Government of Mongolia 2001b: 210). More specifically, the Ministry estimated that 298 boarding schools were in urgent need of repair, and demanded that international donors fund the rehabilitation of at least 193 of them (MOSTEC 2001). In 2002, the Ministry of Education, dismayed by the fact that the international donor community had still not recognized the urgent need for action, commissioned a video documentary in which the dire state of each and every boarding school was made visible to a broader audience. In 2003, after a long decade of complete neglect, international donors, notably JICA (Japanese International Cooperation Agency), the Nordic Development Fund, and ADB (Asian Development Bank) finally agreed to pay more attention to rural school development. In the section "lessons learned," the 3 donors of the second, $63.8 million education sector loan acknowledge the shortcomings of the first Education Sector Development Project (ESDP) with a brief comment: "[T]he Education Sector Development Project and other interventions have . . . revealed the need to include rural areas in projects" (ADB 2002: 10). The Second Education Development Project (2003–2007) funds the rehabilitation of 100 schools in poorer rural areas (half of the initial amount of schools requested by the Ministry of Education) with the aim to "improve learning environments, allow year-round operation, reduce expenditures in energy costs, and ensure acceptable living conditions in school dormitories" (Government of Mongolia and ADB 2002: 42).

This positive turning point might be too little too late. The lack of boarding schools in many rural districts and provincial capitals caused not only the dropout of children from poor nomadic families; it also generated, as mentioned before, a socially stratified system with the potential that boarding schools become stigmatized as second-class institutions for children of families who lacked the resources for better options.

In chapter 5 we noted that international donors, and in particular international financial institutions, had come under attack for imposing their reform packages, and for dictating how governments should improve their economies and educational sectors. As a result of these widespread criticisms, code switching occurred. In the new era of international assistance, international donors learned to speak the language of "national ownership," "sector strategies," "local capacity building," "sustainability," "civil society involvement," and "donor coordination." The example of rural school development in Mongolia, strongly promoted or "owned" by the government itself (especially by the MPRP-led government), elucidates the gap between rhetoric and reality. What the government demanded from international donors—financial support for the rehabilitation of rural schools and boarding schools—fell on deaf ears.

The Statistical Eradication of Dropouts

The argument for entertaining the expensive system of boarding schools has rested all along on a commitment to provide children of nomadic pastoralists with equal access to education. In the same vein, the expectation was that a neglect of boarding schools would produce dropouts in rural areas. There is no doubt in anyone's mind that the boarding school system is in poor shape, yet there is no evidence that the lack of access in rural areas has generated a massive dropout problem. The official statistics on dropouts suggest that access to education is only an issue for a few remote rural areas, not a concern for the country as a whole. As a result, the reform pressure for rehabilitating rural schools and boarding schools is lifted, and there are no immediate incentives for the government to treat rural school development as a priority in the near future. After all, official statistics report a marginal dropout rate of 1.9 percent for school year 2004–2005. An observer might ask: Why bother improving access to education in rural areas? Such a question implies that there is no relation between access to education and non-enrollment or dropout rates. One would qualify as a fool, and rightly so, for reaching such a conclusion.

We propose to take a different angle and call for a closer scrutiny of the politics and economics of official statistics. Official statistics in Mongolia are unreliable sources of information, and yet have far-reaching effects. The fabrication of educational statistics has a direct impact on loans and grants, and government priorities. The problem of false official statistics is of such great magnitude that it deserves detailed exploration. We use the example of dropout statistics as a particular case to illustrate the general tendency to manufacture educational statistics in ways that suit a larger political agenda. However, it would be wrong to assume that government officials are fully responsible for inventing data. More frequently than not, the lead researchers are international consultants who provide technical assistance in developing indicators and subsequently, for collecting, analyzing, and interpreting data.

The prototype of an international consultant operating in Mongolia is a tragic figure in its own right. The international consultants hired by international financial institutions to review, evaluate, or analyze aspects of the educational sector are typically economists. While other international donors are more diverse with regard to professional background, the pattern of employing topic rather than country-experts remains the same. Usually "on mission" for only 1 or 2 weeks, devoid of any country-specific knowledge, fixated on "best practices" that have worked elsewhere, and ill-prepared by a Mongolian counterpart, the consultant clings to countless technical reports produced by his predecessors. Once the joint venture is completed, erroneous statistical information gets circulated within the large international community of development specialists at breathtaking speed.

If we were to believe official educational statistics, the dropout issue is an anathema in Mongolia. In 2004 we came across a vexing phenomenon that unfolded over the course of our evaluation of the Rural School Development Project (Steiner-Khamsi, Stolpe, and Gerelmaa 2004a). Almost all of the 40 project schools reported a significant decrease in dropouts, and many further stated that their dropout problem had been completely eradicated. The steady decrease in dropouts has been purported in official statistics over the past 15 years, not only in project schools of the Rural School Development Project, but in all schools. The official statistics on dropouts tell an incredible story of continuous progress suggesting that the percentage of school-aged children that dropped out of school drastically decreased from an all-time peak of 8.8 percent in 1992–1993, to an all-time low of 1.9 percent in 2004–2005. Table 9.2 presents the official statistics on dropouts.

We soon realized that there is a tacit acknowledgment between practitioners and researchers in Mongolia that the official figures are severely skewed, and

Table 9.2 Official Statistics on Dropouts, School Year 1991/92 to 2004/05 (Definition: Children of age 8–15 that are not enrolled in school)

School Year	Dropouts	Total Student Population	Percentage of Dropouts
1991/1992	33,530	411,696	8.1
1992/1993	33,886	384,069	8.8
1993/1994	23,073	370,302	6.2
1994/1995	16,346	381,204	4.3
1995/1996	14,272	403,847	3.5
1996/1997	16,095	418,293	3.8
1997/1998	14,804	435,061	3.4
1998/1999	15,053	447,121	3.4
1999/2000	13,696	470,038	2.9
2000/2001	13,751	494,554	2.8
2001/2002	13,730	510,291	2.7
2002/2003	11,426	527,931	2.2
2003/2004	11,953	538,398	2.2
2004/2005	10,770	557,346	1.9

Source: MOECS 2005.

the story of ever-decreasing dropout rates is not to be trusted. Our interviewees in the provinces Uvs, Bayan-Ölgii, Khovd, Övörkhangai, Arkhangai, and Bayankhongor were eager to share with us how they counted and registered dropouts. To some extent, their accounts unveiled the mystery of why the official figures for dropouts are so unbelievably low (Steiner-Khamsi, Stolpe, and Gerelmaa 2004a: 86f.).

> Focus Group Interview with 5 Nonformal Education Teachers:
>
> I am not sure whether the dropout rates have really decreased. We still have as many dropouts in our classes as we used to have. Many *sum*-governors nowadays check the box "the family moved away." This way, many children that are dropouts are not registered at all. Our current government proclaims that it resolved the dropout problem, which the previous government supposedly caused.
>
> Meeting with 17 School Teachers:
>
> We reduced dropouts from 172 to 10 students by enrolling them in nonformal education classes.
>
> Interview with Principal:
>
> There are only two school-aged children in our rural district who do not attend school. But they do not count as dropouts. One of them is currently on a leave of absence due to a physical injury [and will return], and the other one has been diagnosed with mental retardation. I checked the medical record of that child myself.

Our exploration into definitions of dropouts revealed that any dropout who is "brought back to school" disappears from the statistics even though most of the nonformal education courses for dropouts only last 2–3 weeks. Furthermore, many local government officials do not bother tracing absentee students, and eventually indicate in their report cards that the family has moved away. Finally, sick children or children with disabilities, however vaguely and disparagingly defined, are excluded from the statistics. We estimate that at least 10,000 children are not enrolled in school because they are physically or mentally impaired.[8] Arguably, the number of these children is too large to be neglected in the official statistics on dropouts (children that prematurely quit school) or "left-outs" (children that never enrolled in school). As the following interview excerpt illustrates, the insensitivity toward children with learning difficulties and poor children is disturbing.

Focus group interview with five teachers and the principal:

> A: We have eradicated the dropout problem in our rural district. 25 dropouts, mostly retarded [direct translation], attended a two-week summer training. At the end, they couldn't read and write, because they were retarded.
> Q: Do you also have other, "normal" dropouts?
> A: In 2000, we had 73 dropouts. We managed to bring 50 of them back to school, and enrolled them in nonformal education. We let them graduate even though they haven't learned much.

Q: Any current dropouts?
A: This year we had six dropouts: three dropped out for health-related reasons, three will come back. The last group of dropouts has to do with their parents. They took them out of school for one or two years to wait until the younger siblings enroll in school. Once the younger children are of school age, they will send the older children back to school. This way, the older child doesn't feel homesick, and can look after the younger siblings. It is very common in our school that ten-year-olds are enrolled at the same time as their eight-year-old brother or sister.
Q: What does your school do for poor children?
A: Sometimes we organize concerts in which poor children perform. Some of that income goes to poor children.

The teachers and principals are not the only ones who blame the students or their parents for dropping out of school. The *Education for All assessment 2000*, for example, shows a similar pattern of explanations that surprisingly went unchallenged by the international community (MOSTEC, UNDP, UNESCO et al. 2000). As the coauthors, four UN organizations (UNDP, UNESCO, UNICEF, UNFPA) and the World Bank approved a report that is saturated with erroneous data, and based upon poor methodology. For example, based on a sample of only 108 dropouts, the Education for All report identified the following reasons for dropout: "to help parents" (38.9 percent), "lack of interest in learning" (31.5 percent), and "health problems" (15.7 percent) (MOSTEC, UNDP, UNESCO et al. 2000: 43; see also Stolpe 2001).

The arbitrariness of explanations for dropouts, and the vast differences in reporting dropouts becomes apparent when we compare official reports with reports of other international and national organizations. For example, the *Adolescent needs assessment report* (Erüül Mendiin Yaam and UNDP 2000) conservatively estimates that there are 4,000 street children, the majority of them in Ulaanbaatar. Yet the Ministry of Education (MOECS 2003b) pledges that there were only 403 dropouts in Ulaanbaatar in school year 2001–2002. If we were to believe the official statistics, only 80 girls in Ulaanbaatar, and 2 girls in the entire province of Khovd, dropped out for poverty-related reasons. Although the population of Ulaanbaatar accounts for one-third to half of the country's population (depending on the information source), the city has supposedly three times fewer dropouts than Arkhangai province. The only credible information in the official statistics is the ratio between female and male dropouts which corresponds to a general pattern of females being enrolled in greater numbers and for a longer period of time than males. The inverse gender gap, discussed in chapter 6, is also manifest in the dropout ratios.

The saga of official dropout statistics would take an entire volume to document, and would fit well in a series of other books that deconstruct distortions such as *How to lie with statistics* (Huff 1954), *How to lie with maps* (Monmonier 1991), or *How to lie with charts* (Jones 1995). Thus far, we have commented on how official statistics underreport dropouts and thereby perpetuate the myth that the dropout problem, after a peak in the early 1990s, is slowly but surely subsiding. However, this type of "statistical eradication" of dropouts from official records is only half of the story. The other half is even more intriguing for a study on borrowing and bending. There are two phenomena, in particular, that deserve mention: the mismatch of figures provided by different government sources and the retrospective fabrication of dropout figures.

First, the *Mongolian Dropout Study* of the Mongolian Education Alliance (del Rosario, Battsetseg, Bayartsetseg et al. 2005) assembled dropout statistics for school year 2003–2004 from four different sources. Although all of them are government agencies, their figures vary considerably. The difference between the figures provided by the Ministry of Education/National Statistical Office (11,953 dropouts) and the Human Rights Commission of Mongolia (68,115) is by far the largest. Surprisingly, there is also a big difference between two departments within the same ministry: Whereas the Nonformal Education Department reports 40,000 dropouts, the department in charge of educational statistics at the Ministry of Education only reports, in line with the National Statistical Office, 11,953 dropouts. Table 9.3 presents the comparison of four government agencies that each manufactures different dropout statistics.

The vast discrepancies in reporting dropouts warrant interpretation. The various government agencies appear to send different signals to different constituencies. Whereas the Ministry of Education and the National Statistical Office are accountable to the general Mongolian public as well as to international financial institutions that subsidized educational reform over the past 15 years, the Nonformal Education Department at the Ministry of Education, together with UNICEF, is seeking additional funds for financing nonformal education programs for dropouts. Their price tag must reflect the large number of potential participants (40,000) who would be served in nonformal education programs. In stark contrast, the Government of Mongolia is under pressure to provide evidence that the reforms of the past 15 years have been effective in eradicating the dropout problem.

Table 9.3 Comparative Figures on Dropout Rates—School Year 2003/04

Agency	Dropout Statistics	Difference from Highest Figures
Human Rights Commission	68,115	—
UNICEF and Nonformal Education Department of Ministry of Education	40,000	−28,115
Census 2003	17,671	−50,444
Ministry of Education and National Statistical Office	11,953	−56,162

Source: del Rosario, Battsetseg, Bayartsetseg et al. 2005: 44.

Second, the business of mythmaking does not stop short at retroactively tampering with educational statistics. In 1993, the Government of Mongolia stated the following:

> Dropout rates have increased from 4 percent in 1988/89 to almost 22 percent in 1992/93, with those in rural schools (especially males) being the more common dropouts. (Government of Mongolia 1993: vi)

The high dropout figure of 22 percent generated an international appeal for financial support. The signal worked. Based on this horrific number of dropouts (every fifth student), international donors (notably DANIDA) immediately funded large-scale, nonformal education programs for dropouts. Similarly, panic-stricken by the rising levels of poverty, huge dropout rates, and the dramatic decline of literacy rates in the beginning of the 1990s, the National Poverty Alleviation Program was formulated in 1994, and implemented two years later (Government of Mongolia, World Bank, and UNDP 1999: section 5.2).[9] Mongolian teachers have vivid memories of the mid-1990s when dropouts and left-outs were brought back in droves for nonformal education classes, which were offered during school vacations or on Saturdays. Although the memories remain, the numerical evidence vanished from government records once the loans and grants were received. Seven years later, the 2000 *Education for All assessment* (MOSTEC, UNDP, UNESCO et al. 2000) maintains that there were only 8.8 percent dropouts in 1992–1993, not 22 percent as previously suggested. It has become a common practice among government officials to resort to methodological explanations when discrepancies in official statistics surface. They argue that the previous indicators yielded unreliable data, and were phased out as a result. Not surprisingly, the new indicators generate data that describe a story of recuperation from the shock therapy of the early 1990s, and portray gradual progress since 1995. The replacement of indicators from one dropout study to another has had several precedents in other fields, notably the shift from income- to consumption-based indicators for measuring poverty (National Statistical Office of Mongolia and World Bank 2001). Even with a wild imagination, one is hard-pressed to believe that the distortion of dropout rates is based solely on methodological error. Nevertheless, beginning with the Education for All assessment in 2000, the government has grossly underreported figures for dropouts, and the reports of international consultants have uncritically perpetuated them.

GLOBAL ENCOUNTERS WITH LOCAL NEEDS

Many educational researchers have marvelled at the speed with which specific reform strategies such as outcomes-based education, decentralization, or vouchers, have spread around the globe. Not surprisingly, the question of whether the global diffusion process also follows the same critical stages as epidemics has invigorated the imagination of researchers. In contrast to studies on how global contagions in educational reform actually work, little has been said about best practices that evaporate and cease to be systematically pursued, let alone disseminated.

The example of nomadic education in Mongolia is an interesting case of a transfer vacuum that led to a stalemate situation: the government neither received (financial) support from international donors for transferring positive experiences from the socialist past into the postsocialist present, nor was it pressured by them to import, selectively adopt, or experiment with alternative models of formal nomadic education that had been implemented in other educational systems. In fact, with the exception of DANIDA, international donors were utterly disinterested in offering a solution for rural school development in Mongolia.[10] They identified the situation in boarding schools as an internal "problem," because international donors could not agree on "best practices" in nomadic education that would work in Mongolia. At a loss as to what to recommend or impose, they abstained from any action throughout the long decade of neglect. The case study of nomadic education has made us aware of the lack of studies which examine reforms that do *not* travel, neither from the past to the present, nor from one education system to another. Most scholars in globalization studies investigate local responses to global forces (e.g., Carnoy and Rhoten 2002; Anderson-Levitt 2003) by following the routes of traveling reforms. We propose to examine the inverse as well: how international donors react to local solutions. Drawing on the example of nomadic education in Mongolia, we found that international donors disregard "national ownership" or local forces, and dismiss locally developed solutions.

We conclude the case studies on imposed, voluntary, and no borrowing with a comment on globalization studies. As previously suggested, there is a peculiar asymmetry in globalization research that became apparent when we analyzed the long decade of neglect in nomadic education. The focus of globalization research is almost exclusively on local encounters with global forces, and rarely the other way around; it is erroneously assumed that there are no local forces. However, as this chapter demonstrates, there are many insights to gain from studying how global players encounter local forces. Such studies would elucidate how low-income governments depend on international donors transferring funds, visions, and "best practices" from one country to another.

10

BENDING AND BORROWING IN MONGOLIA, AND BEYOND

Globalization is commonly viewed as an act of de-territorialization (see Appadurai 1990). By implication, globalization studies investigate the flow of money, communication, beliefs, or—as is the case in this book—the travel of educational reforms from one cultural context to another. Such a research endeavor is more ambitious than it appears. The greatest challenge is to avoid falling into the trap of first establishing national boundaries, only to demonstrate afterwards, that these boundaries have indeed been transcended. Reforms do not have a home base, a territory, or a nationality, and therefore do not "belong" to a particular educational system. Individuals conceive reforms and, depending on where they are geographically and institutionally situated and how well they are globally networked, succeed in having their ideas disseminated worldwide. This book is not about global policy entrepreneurs or policy networks that carry some ideas forward at the expense of others. Rather this book is about understanding why policy makers in one country—Mongolia—*refer to* globalization, that is, generate reform pressure by pointing at educational reforms in other countries.

We have taken a stance that is opposed to convergence theorists: Although on the surface it appears that educational reforms in Mongolia follow the same pattern as in many other countries, the similarities disappear once a reform is examined on-site. Indeed, if we were to only listen to how government officials and international donors speak about developments in Mongolia, we would be duped into believing that all global market-oriented reforms, ranging from the standards to the decentralization movements, also made it to Mongolia. But a closer look reveals that policy borrowing in Mongolia occurs either rhetorically or selectively with limited impact on existing practices. And yet, politicians, policy makers, and educators in Mongolia insist that educational reform in their country follows the same international standards as in other countries. There is obviously a message embedded in these types of public announcements that we have sought to decipher in this book.

Arguably, the convergence question is too broad to yield new insights. It is flawed because it does not account for the existence of different policy levels. Of course, there is *no* convergence at the level of policy implementation;

after all, contexts vary, and policies play out differently in various cultures. Furthermore, no convergence believer seriously claims that reforms are enacted in the same way, generating similar legislations and policy guidelines. There is always and everywhere a huge gap between policy talk and policy action. Stressing the loose coupling between envisioned and enacted policies is therefore also a moot point. What are we left with when we take the distinction between policy talk, policy action, and policy implementation into account? At what level is convergence supposed to occur? Perhaps, what globalization does to national educational reforms is not more, but also not less, than propelling brand name piracy whereby every government borrows the same label, but gives it an entirely different meaning. Once we acknowledge the "global speak" of government officials, we begin to pay attention to the benefits of using a universal language of educational reform. Why this insistence on being part of a global reform movement? What is there to gain from aligning educational development with imaginary "international standards" in education? These kinds of questions immersed us in the politics and economics of educational borrowing in Mongolia. By way of highlighting the main points made in the previous chapters, we offer here a few propositions on how the study of imported reforms in Mongolia might be used to inform the larger fields of globalization and policy studies in education.

Educational Transfer and the Study of Globalization

Chapter 1 introduces important background information, and explains why we find the case study of Mongolia suitable for examining the process of globalization. Mongolia is a case of a late adopter in that it borrowed reforms that had been traveling around the globe for quite sometime. This special feature of policy borrowing makes Mongolia especially compelling for globalization research. Rather than reiterating our interpretive and methodological framework, presented in chapter 1, we proceed with identifying more explicitly the conceptual ties between borrowing and globalization research.

The revived interest in borrowing studies has challenged researchers in comparative education to be more specific in what they think they have to offer to the study of globalization. Ironically, most scholars in comparative education see themselves as borrowing researchers even though two distinct transnational interactions are involved: borrowing and lending, import and export, or reception and diffusion. We believe it is more accurate to investigate the two interactions separately, and subsume them under the label "educational transfer." Conversely, we briefly sketch how educational transfer research was transformed in light of the globalization challenge, and highlight a few contributions and shortcomings of transfer research.

Comparative research on *policy borrowing* underwent several major discursive shifts. An important one was the move from normative to analytical studies—the first being concerned with what *could* and *should* be borrowed, and the latter interested in understanding *why* and *how* references were made to experiences from elsewhere. Jürgen Schriewer (1990) must be credited for criticizing normative and meliorative approaches to the study of policy borrowing. Embedded in a theoretical framework of system theory (Luhmann 1990), Schriewer and his colleagues propose to study the local context in order to understand the "socio-logic" (Schriewer and Martinez 2004: 33; see also Schriewer, Henze, Wichmann et al. 1998) of externalization. According to this theory, references to other educational systems serve as leverage to carry out reforms that otherwise would be contested. Schriewer and Martinez (2004) also find it indicative of the "socio-logic" of a system that only specific educational systems are used as external sources of authority. Which systems are used as "reference societies" (Schriewer, Henze, Wichmann et al. 1998: 42) and which aren't, tells us something about the interrelations of actors within various world-systems. Pursuing an analytical rather than a normative approach to the study of educational borrowing, one reaches a conclusion contrary to what borrowing advocates have asserted: Borrowing does not occur because the reforms from elsewhere are better, but because the very act of borrowing has a salutary effect on domestic policy conflict.

We applied the concept of externalization to comparative policy studies, and found that it is precisely at a moment of heightened policy contestation that references to other educational systems are made. Thus, borrowing, discursive or factual, has a certification effect on domestic policy talk (Steiner-Khamsi 2004b). Against this backdrop of system theory three common phenomena, which at first appear to be nonsensical, make perfect sense: (1) very often the language of the reform is borrowed, but not the actual reform, (2) borrowing occurs even when there is no apparent need, that is, even when similar reforms already exist in the local context, and (3) if the actual reform is borrowed, it is always selectively borrowed and sometimes locally recontextualized to the extent that there is little similarity left between the original that was emulated, and the implemented reform.

The concept of externalization has also been applied to *policy lending*, and political and economic reasons for policy export have been examined. Phillip Jones (2004), for example, focuses on the dual meaning of the World Bank's portfolio in education: the Bank's portfolio with regard to *loans* and its portfolio with regard to the *lending* of ideas about educational reform in low-income countries. Although using finance as a means to drive policy change is hardly new, the scale and global reach of international organizations prompts key questions of substantial interest for education theory, policy, and practice (see Boli and Thomas 1999; Chabbott 2003). Many scholars have criticized transnational regimes, particularly the World Bank, regional banks including the Asian Development Bank, UN organizations, and nongovernmental organizations for making their loans or grants contingent on whether governments agree to sign off on the import of specific reform packages. Such economic contingencies have led politicians to "speak" in one manner, and act in another. In effect, the transnational regimes encourage brand name piracy or discursive borrowing of internationally renowned concepts such as "vouchers," "outcomes-based education," or "student-centered learning," and help to catapult them from one corner of the world to the other.

The "strings attached" to a grant or loan (contingency) along with organizational concerns (cost-effectiveness, manageability, visibility) render the double-talk endemic to international cooperation projects. From the perspective of international organizations, the transfer cost—the cost of local adaptation to make a borrowed model effective—is lower than having high-paid international consultants supervise projects that were locally developed. Besides concerns for cost-effectiveness and manageability, visibility is yet another consideration. In an environment of competing funding and ever-enlarging portfolios, both between and within international organizations, it is of utmost importance that projects are "correctly" identified as belonging to the organization that has funded them. This applies, in particular, to organizations such as UNICEF or UNESCO that are constantly struggling with funds, and depend on remaining visible despite little financial support. Of course, the three concerns for cost-effectiveness, manageability, and visibility, accounting for the global dissemination of institutional "best practices," have different implications for various international organizations.[1] Nevertheless, they are three of the most common reasons why international organizations or transnational regimes promote their own reform packages in the countries where they are operating or governing.

Transnational Interaction and Cooperation from a Historical Perspective

In chapters 2, 3, and 4, we deliberately move beyond the bias in globalization studies that focus exclusively on developments in the twentieth century. Educational import in Mongolia did not begin in the 1990s. What has transpired from educational practices in other countries over the course of three centuries is essential for understanding the contemporary Mongolian institution of schooling. Historical accounts of external influences on educational development in Mongolia are at best sketchy, and virtually nonexistent in English. We have therefore taken on the task of analyzing cultural encounters in Mongolia that left deep traces on the educational system.

In chapter 2, we present four of the five periods of educational import: enlightenment, colonization, nation building, and universal access to education. The reference societies, from where educational concepts were either voluntarily borrowed or forcefully imposed, varied across time. They include Tibet, Manchuria, European countries, and, during the final period, the Soviet Union. The other chapters of this book, particularly chapters 5–9, deal with the fifth period, beginning in 1990, when policy makers borrowed educational reforms—rhetorically or selectively—from wherever international funding was made available.

We have used an extended version of world-systems theory to demonstrate that postsocialist Mongolia inhabits two different world-systems simultaneously. Economically, it is considered a peripheral country that depends on external bilateral assistance from established market economies, especially from Japan, Denmark, Germany, the United States, and the Republic of Korea, as well as from international financial institutions and multilateral donors. These donors treat Mongolia the same as any other developing country. Politically, however, Mongolia shares the same transnational "space" (Nóvoa and Lawn 2002) with over 30 former socialist countries. This statement is not as banal as it may sound given that, for bureaucratic reasons, the United Nations and the World Bank have mistakenly assigned Mongolia to the Asia-Pacific region.[2]

The commonality of reforms in postsocialist countries, especially in Central Asia and Mongolia, has several causes. The educational systems in these countries were until 1990, with a few exceptions, almost identical, reflecting Soviet influence in the region. Moreover, these countries not only experienced the structural reform policies during the same time period (early and mid-1990s), but these policies were administered by the same international donors.

For another research project we gathered some of the features of the "postsocialist reform package" that was transferred to countries in the Caucasus (Armenia, Azerbaijan, Georgia), Central Asia (in particular, Kazakhstan, Kyrgyzstan, Tajikistan, and Uzbekistan), and Mongolia (Silova and Steiner-Khamsi 2005). Apart from Tajikistan that had a "late start" (1997) with receiving international loans and grants due to civil unrest, governments in every other country in the region received funding to implement the following package:

- Extension of the curriculum to 11 or 12 years of schooling
- Introduction of standards and/or outcomes-based education
- Decentralization of educational finance and governance
- Reorganization of schools ("rationalization" of staff and structures)
- Privatization of higher education
- Standardization of student assessment
- Textbook reform
- Establishment of education management and information systems

Each of these components of the reform package attempted similar implementation strategies. For example, the goal of decentralizing educational finance and governance—pursued with mixed outcomes—made it necessary to build management capacity among school administrators and educational authorities. As a consequence, every country in the region experienced a proliferation of professional development courses and, in some cases, certification programs on school finance, management, and administration. Arguably, what was stressed in the reform package is as interesting as what was omitted. In each case there was limited support from international financial institutions for preschool education, and virtually no support for students with special needs. Moreover, the reforms in these countries were generally unsuited for rural schools, and accelerated urban-rural differences.

The educational reforms in Mongolia are—despite the same Soviet-style educational system in the past, and an identical package of reform in the present—quite unique. As the case studies in this book illustrate, the mere existence of a reform package does not necessarily imply that the reforms are ever fully implemented, or even implemented at all.

LOST IN TRANSITION

Ironically, there is much more written about the alleged end of history (Fukuyama 1993) than about its beginning. Reading the international literature on educational development in Mongolia, however, evokes the image of a clock that was reset to "year zero" in 1990. What counts most, judging from international authors writing about Mongolia, is what happened *after* 1990. A few authors also move beyond the transition framework and, in passing, mention the last couple of years *before* the Big Bang in Mongolia; if for no other purpose than to make the contrast between socialist and postsocialist education more glaring. Regardless of the exact reasons—language barriers, difficult access to literature, convenience—that account for such a shortsighted perspective, the consequences for educational reform in Mongolia are far-reaching. The shortsightedness is especially pronounced with international experts who prepare technical reports and reviews on behalf of international donors in Mongolia. Lacking a thorough understanding of the history of education in Mongolia leads them to make false claims, such as that all educational reforms in Mongolia after 1990 are new, better, and global in orientation. If, against all odds, the reforms turn out to be less promising and more chaotic than anticipated, there is a ready-made answer: "transition."

Forgetting the past, and restarting history is not out of the ordinary in countries that have undergone massive political changes (see Foner 2002). The twin concepts—"annihilation of the past" and "reversibility of events"—put forward by Christian Giordano and Dobrinka Kostova (2002: 77), poignantly capture how the history of Mongolia and other former socialist countries has been reinvented twice in the twentieth century. After the revolutionary changes of 1921, historians generated a socialist interpretation of Mongolia's past. These accounts were invalidated after 1990, and replaced with an anti- or pre-socialist interpretation of the history of Mongolia. Giordano and Kostova explain,

> The former [annihilation of the past] refers to the systematic elimination of past facts, symbols and social practices that are believed to be the heritage of eras labeled "barbaric," "obscurantist" or "degenerate." The latter [reversibility of events] refers to the project of restoring matters "as they were before," thus leaving behind a recent past now exposed as a fatal mistake. (Giordano and Kostova 2002: 77f.)

The two methods of historiography, applied after each revolutionary change, differ in that the socialist narrative of history was based on the "past's selective destruction" (Giordano and Kostova 2002: 78), whereas postsocialist history attempts to present the events of the past 70 years as reversible. As the authors succinctly summarize,

> One narrative [socialist history] is "prospective" while the other is "retrospective," implying that the postsocialist future can begin by returning to a *status quo ante*. According to the logic of this presentation, it is necessary and desirable to recreate the conditions of the pre-socialist era as if socialism had never existed—or as if it existed only outside the "correct flow of history." (Giordano and Kostova 2002: 78)

In another fascinating study, Augusta Dimou (2004) notices the same retrospective narrative in contemporary German history textbooks. The history of the Soviet Union from 1920 to 1980 is written, albeit very briefly, in a chronological order of important events. In fact, chronology is a key feature of history textbooks. In stark contrast, the history of the Soviet Union and communism in the 1980s and 1990s is narrated retrospectively, starting out with a detailed portrayal of the fall of communism, and then moving back in time to interpret the milestones in the 1980s that must have, from today's perspective, paved the path for the demise. The retrospective narrative employs a "teleology of defeat" (Dimou 2004: 353) in that the collapse of communism is first locked in as the main reference point, and all the historical events before "year zero" are spun into a causal web that supposedly explains why the defeat was ultimately inevitable. It was important to us not to fall into the same trap in the chapters in which we touch upon the history of education in Mongolia.

Inspired by anthropologists and other social scientists who have written in the newly emerging field of postsocialist studies (e.g., Hann 2002), we use a time frame that acknowledges important developments before the 1990s. Additionally, we purposefully refrain from the contrastive method that superficially juxtaposes the educational reforms in the postsocialist era with the last couple of years before the collapse in 1990. Instead, our historical accounts attempt to illuminates the entire socialist period 1921–1990, and the periods of educational import before 1921. We find a long-term perspective indispensable for examining the continuities from the past to the present, and to understand the cultural context of globalization in Mongolia.

Chapter 3 explains the Marxist-Leninist "bypassing capitalism" narrative, which enabled Mongolia to reinvent itself as a nation that, having successfully left the chains of "feudal oppression" behind, skipped the stage of capitalism, and moved, after a long period of transition, into the socialist stage. The end goal (suspended in 1990) was communism. The chapter presents a few of the far-fetched ideological constructions fabricated in order to bring the atypical Mongolian context in line with the Marxist-Leninist theory of development stages. Originally written for the exploited working class in industrialized nations, the theory was a bad fit for Mongolia. The necessary ingredients for a revolution, such as a clear distinction between classes, had to be first assembled and then adapted to the Mongolian context. These "facts" calling for revolutionary changes weren't constructed once and for all, but were periodically invoked, placing Mongolia in a status of perpetual "transition."

The first transition period lasted 19 years. It began with the revolution in 1921 and ended with the Tenth Party Congress of the Mongolian Revolutionary People's Party in 1940, where the "creation of the foundations for the development of socialism" (*sotsializmyn ündsiig baiguulakh üye*) was inaugurated. Having embraced a people's revolution, Mongolia then had to successfully complete a phase of "transition," purging all legacies of the feudal past, before entering the stage of socialism.[3] According to Marxist-Leninist historiography, this "revolutionary democractic phase" lasted in Mongolia from 1921 to 1940. The "transition" from feudalism to socialism in Mongolia coincided with the Stalinist period of violent purges, brutal persecutions, and failed attempts at collectivization. The government of the Mongolian People's Republic eliminated "elements" that were "counterrevolutionary" or feudal, including intellectual lamas and aristocrats. Finally in 1940, after almost two decades of "anti-feudal" struggle, the country was prepared to enter the stage of state-socialism. In effect, the plan of the Tenth Party Congress to generate the "foundations of socialism" in 1940 meant that a transitional phase was once again underway.

At this point the two most visible markers of socialism—an "industry" and "the working class"—had to be "created" (see Tüdev 1971; Nansal 1971; Shagdarsüren 1976). However, industrialization was difficult in an economy that relied almost exclusively on nomadic pastoralism. It was taken for granted that building the required human capacity and manpower—a role assigned to the educational system—would stretch over a long period of time. Culture campaigns were another attempt to purge the "socialist man" of the illiteracy and immorality of the dark ages, and prepare citizens for the move from an agrarian to an agrarian-industrial society. This second stage of transition came to an end in the 1960s when socialism was considered more or less established. The end of one transition was the beginning of the next.

During the period 1961–1980, the Mongolian, People's Republic entered the stage of "creating the material and technical base of socialism," (Shirendyb 1981: 23; Sükhbaataryn neremjit khevleliin kombinat 1964: 45) with the goal of transforming Mongolia from an "agrarian-industrial into an industrial-agrarian country" (Pelzhee 1981). From then on, whatever the new epochs were called, Mongolia found itself alongside every other socialist country in a never-ending state of transition (Mongolian: *shiljiltiin üye*). This was manifested everywhere in the infinite absence of the promised communist paradise. In all socialist states during this period of permanent transition, the absence was justified by the invention of ever-new substages. Ironically, Mongolia's state of never-ending transition has been carried over from the socialist past to the postsocialist present.

This brief account of Mongolia's socialist "transition" periods allows us to reflect on its last, which began in 1990. Several scholars note how socialist practices from the past have endured in the postsocialist present, manifesting themselves in "the more straightforward infra-structural legacies such as the administrative-bureaucratic legacy, and the more elusive political cultural continuities" (Barkey and von Hagen 1997: 188). However, there is a paucity of literature on how postsocialist governments, for a variety of reasons, signal the beginning of a new era by deliberately distancing themselves from the socialist past, and proclaiming programs that intuitively come across as "antisocialist." Arguably, such was the case with several educational reforms of the past decade in which Mongolian politicians and government officials professed to pursue a new (nonsocialist) path in the "age of the market" (Sneath 2002a). For research on globalization in Mongolia, the "transition" status is essential for understanding why so many traveling reforms were only borrowed rhetorically, and never implemented. For any policy maker in Mongolia it is abundantly clear that borrowing educational reforms from elsewhere is a prerequisite for admittance to the international community of established market economies, thereby ending the stigma of transition.

The important role of education in accelerating social change in Mongolia cannot be overstated. During the socialist period, alphabetization and universal enrollment was one of the key indicators of progress; so much so that the revolution of 1921 was falsely credited for enrollment figures of the 1920s, and adult literacy rates were periodically inflated in the 1950s and early 1960s: At the end of the First Five-Year Plan (1948–1952) the achievement of universal literacy was prematurely proclaimed and celebrated (Mandel 1949). The proclamation was revoked ten years later, when the socialist government embarked on two "culture campaigns" (1960–1961 and 1962–1963) to eradicate illiteracy, alcoholism, epidemics, and vandalism. Today, although the international financial institutions give low priority to the social sector,[4] Mongolians highly value formal education, and evaluate action programs of political parties in terms of what they have to offer or promise, with regard to child subsidies, preschools, and schools. Education is a highly politicized issue in Mongolia, placing government officials in an awkward position where they must please two divergent constituencies. On one hand, they are forced by international financial institutions to "deregulate" and "decentralize" education in the age of the market, and, on the other hand, they have to prove to an education-minded populace that education is on the top of the government agenda. This tension explains, to some degree, why dropout rates in Mongolia have been (statistically) eradicated, and why educational statistics are manipulated to demonstrate progress in educational development.

LIFTING THE IRON CURTAIN IN EDUCATIONAL RESEARCH

Chapter 4 explores "donor logic" in Mongolia, both from a broader historical perspective, as well as a narrower focus on international donors currently involved in the country. For more than half a century Mongolia has been very dependent on external financial sources to operate and reform the educational system. From 1962 to 1991, funds for the educational sectors were made available by CMEA, and beginning in 1991 by international financial institutions (especially ADB), bilateral donors, UN organizations, and international nongovernmental organizations, such as the Mongolian Foundation for Open Society,[5] Save the Children UK, World Vision, or the Danish Mongolian Society. The diversity of external funding sources that replaced the single source (CMEA) in 1991 is remarkable. However the total amount provided as a percentage of the GDP is slightly lower, and the ratio allocated to the social sector (including education) is considerably lower, than before 1991. It is uncontested that the priorities of post-1990 international donors shifted from the social sector to the economic sector (World Bank 2004a).

Along with different funding priorities, the shift from internationalist funding (from socialist countries) to international loans and grants (from market economies) also implied a diminished status for Mongolia in world affairs. Even though Mongolia heavily relied on funding from CMEA, it was not officially depicted as an aid-dependent or developing country. The ideology of "fraternalist solidarity" not only employed a different language of assistance, but a different conception of progress. Whereas international financial institutions today lend money for economic growth, financial and political stability, and for global trade, the member states of CMEA vowed to support each other in the common pursuit of the "socialist path" of development. Two dimensions, an economic and an ideological, were involved. For one, the economic aspect implied that poorer countries, such as Mongolia, received resources, technical assistance, and other kinds of support from CMEA, to eventually bring them up to par with other member states. Furthermore, the ideological feature of "fraternalist solidarity" meant that the older socialist systems guided the newcomers in the right direction politically. Regarded as a flagship socialist country that had already embraced a people's revolution in 1921, Mongolia was a place of ideological pilgrimage for the leaders of communist parties in Mozambique, Tanzania, Cuba, and other countries coming to marvel at the "socialist path" Mongolia successfully pursued, especially in education. This particular representation of Mongolia as a country that was economically poor but ideologically respected, is essential for understanding why education in Mongolia was celebrated as one of the most visible markers of (socialist) progress. We have noticed that the status shift from Second World to Third World country and from lender to borrower of ideologies, is severely neglected in the literature on educational development in Mongolia.

It is important to bear in mind that for over 40 years (1949–1991) 2 global "aid" systems coexisted side by side, each serving their own "member states."[6] An "iron curtain" was set up between the parallel universe, preventing the flow of people, capital, communication, and goods across the two world-systems. This ideological divide, which had far wider repercussions for education than is commonly assumed, merits greater attention. Arguably, comparative education journals function as a window for viewing how other educational systems are represented. Clearly the *Comparative Education Review*, published in the United States, paid considerable attention to educational developments in the People's Republic of China and the Soviet Union during the period of the Cold War. An example of the Cold War climate in educational research, is provided by a review of the book *Neoproverzhimoe* (Kuznetsov and Kashoian 1963), published in Moscow.

> Life of the youth in capitalistic countries is described as hard and sad. This book gives the facts concerning the bitter fate of the young generation in the countries of the "free world." These facts express life itself and the cruel reality of the capitalistic "paradise." The youth is the future of the people, but in the "free world" it is deprived of political rights and its first experience in life is unemployment. Education is a privilege of rich people only; for the poor it remains a dream. In the U.S.A. one out of ten is illiterate. The book is written for a popular audience. (Lipski 1963: 95)

What's worth noting about this book review is not only the vilification of the U.S. educational system by the Russian authors, but also the tenor in which books from the Soviet Union or the People's Republic of China were reviewed in the *Comparative Education Review*. Several summaries by the reviewer Vladimir Lipski end with the assessment, "[T]his book is written for a popular audience" (Lipski 1963: 95), encouraging the reader to dismiss any educational research book published in the Soviet Union as communist propaganda.

The Iron Curtain separating the two world-systems until the mid-1980s, has not yet been lifted in educational research. Delving into an analysis of the Cold War period—especially its height in the 1950s and 1960s—is especially promising for understanding how domestic reforms have been shaped by an absent and threatening "other." Such a project could reveal both the salutary and adversary effects that the two equally strong superpowers have had on educational developments in their respective countries. Although the image of the United States as a "racist imperialist superpower" was predominantly circulated within the closed world-system of socialist countries, the accusations periodically leaked, and were advanced by political parties and groups in the West that were sympathetic to communist ideas. Scholars have only now begun to examine whether the constant and vociferous critique by the competing superpowers might have generated reform pressure on public policy domestically. For example, in 2004, on the occasion of the fiftieth anniversary of the Brown versus Board of Education legislation, the question was raised whether the negative perception of the U.S. educational system as racist and oppressive, promoted by the Soviet Union and its allies, had a positive impact on U.S. legislation in education. The example of Brown versus Board of Education, which terminated the de jure discrimination of racial minorities in the U.S. educational system, is but one example of how the existence of two competing superpowers helped reinforce criticism within each system, shaping educational reforms in the two world-systems.

Although research on the Cold War period is in its infancy, we believe that several theories must be revised based on newly accessible archival material. For example, chapter 4 also presents a few theories on the role of international donors, and we propose revisiting the role of the UN system in light of post–Cold War research on these issues. There are fascinating accounts of how the UN system transformed itself from being a Western entity to an organization that gave voice to its numerous member states from the Third World (Jones 1988; Mundy 1999). In contrast, there is not much written about the United Nations as an institution that bridged the two world-systems. The UN system represented one of the few global institutions, if not the only one, that crossed the ideological divide during the Cold War. What we also found striking is that both world-systems—the socialist and the capitalist—laid claim on the UN project of world peace, international understanding, and universal collaboration, and identified their own world-system as the driving force behind the project.

Encounters with Traveling Reforms in Mongolia

In addressing questions that surface at the intersection of globalization and borrowing/lending research, we utilized two themes common to our case studies: What under-explored aspect of globalization research does the case represent, and how relevant is the case for understanding what has been going on in the Mongolian education sector reform over the past one-and-a-half decades? Guided by these questions, it became clear that political and economic reasons for borrowing must be prioritized. This assertion should not come as a surprise since Mongolia was forced to reorient itself in 1990, and profess policies and reforms emanating from its new allies in high-income market economies. By the time Mongolia joined the pool of aid-recipients, the international donors had thick portfolios with "best practices" in place, tested in numerous other postsocialist and low-income countries. The timing of policy borrowing in Mongolia matters a great deal. Studying Mongolia as a late adopter, we discovered educational reforms that had already experienced a global career, notably student-centered learning, vouchers, and outcomes-based education. Similarly, the structural adjustment policies were by no means unique to Mongolia; they were imposed on all countries that borrowed money from international financial institutions. Additionally, by acknowledging that policy borrowers are not helpless victims but active agents in what they selectively borrow and how they modify what they have borrowed, we draw attention to Mongolian responses to imported reforms. Thus, emphasis in this book has not only been on how politicians in Mongolia had to bend to reform pressures from international donors, but also how they subsequently bent the global reforms that they had imported to suit the Mongolian context.

Chapter 5 is a chapter (along with chapters 6 and 7) in which we investigate how existing practices are impacted by policy import. Here we critically examine early reforms that were launched in the early and mid-1990s. Aiming at drastically reducing public spending for education, and at the same time generating income for schools, these reforms revamped the structure and organization of the previous system. Funded by the Asian Development

Bank, and known as structural adjustment policies (SAPs), the story of these radical educational reforms has remained largely untold. We focus on three of these imposed reform packages and analyze what is left of them, ten years later. Rarely do educational imports entirely replace existing practices. We therefore selected the three structural adjustment reforms that to various degrees replaced previous practices: tuition-based higher education, decentralization of educational finance and governance, and rationalization of educational structures and staffing. Whereas a few reforms (tuition-based higher education, abolishment of schools in remote villages, reduction of school staff) were irrevocable, other reforms, as nonnegotiable as they appeared at the beginning of the SAP decade, were eventually dismantled or subverted after a few years. One of the most irrational "rationalization reforms" was the downgrading of schools in rural districts (*sum*). The majority of schools in rural areas were forced to close the two final grades (9 and 10), and place continuing students in regional (inter-*sum*) schools that offered the full range of grades. The reform failed miserably, because parents, teachers, and principals boycotted it in subtle and imaginative ways. The provincial educational authorities were forced to bend under pressure from rural schools, and approved one request after another to reopen grades 9 and 10, thereby undermining the structural adjustment policy, imposed by international financial institutions, and rubber-stamped by the Government of Mongolia.

Chapter 6 deals with hybridization, and it investigates how the import of student-centered learning plays out in a class setting that is both hierarchical and teacher-centered. A key for understanding the social order in the Mongolian classroom is the system of class monitors. Mostly female students, class monitors are peers who serve as assistant teachers, discipliners of students, and organizers of social events. Our attention is drawn to how teachers, class monitors, and regular students have together Mongolized the Western import of student-centered learning in the classroom. In an effort to make sense of how student-centered learning has been indigenized or reinterpreted, we present a few cultural constructions of the "good" teacher and the "good" student that have shaped Mongolian pedagogy.

These strong pedagogical beliefs are counter to the neoinstitutionalist assertion of a world-culture in schooling that supposedly makes teachers in different parts of the world teach similar content in a similar manner. Embedded in a neoinstitutionalist framework, David Baker and Gerald LeTendre (2005) analyzed massive amounts of data from the Third International Science and Mathematics Survey to make the case for a world-institution of schooling, leading to a convergence of pedagogical beliefs and practices. We have not found any evidence that such a claim could be made on behalf of the Mongolian classroom. Even teachers who had been exposed for years to student-centered learning techniques only selected those aspects of the imported "new teaching technology" that fit with their own pedagogical beliefs and practices. William K. Cummings (2003) provides an important historical analysis in which he demonstrates the existence of several "institutions" of schooling that had global influence. Leaning on Cummings' work, we adopt a multi-institutional approach, and suggest that schooling in Mongolia has been shaped more by historical developments in the country and the region than by a world-culture or world-institution of schooling. Since class monitors are mostly female, we have also taken the opportunity to engage an issue that is hotly debated in and outside of Mongolia: the reverse gender gap where Mongolian females outperform males.

Chapter 7 explores how the import of outcomes-based education (OBE) has affected existing practices in Mongolian schools. Introduced in 2003, OBE was the offspring of a general public management reform that swept through all public sectors in Mongolia and was applied to all civil servants, including teachers. The reform is intended to replace interpersonal and informal agreements with measurable performance agreements or contracts. Strikingly, the New Contractualism in Mongolian schools neither replaced existing practices nor hybridized them. Instead, the teacher scorecards or the "outcomes-contracts" merely reinforced the elaborate teacher surveillance system that had already been in place for the past thirty years or so. The system has traditionally been closely linked with monetary rewards and sanctions based on teacher performance. More than 40 percent of a teacher's full income is composed of performance-based salary supplements and bonuses. The opposite, however, also applies. Deductions from salaries and supplements are made; in fact so frequently that they are the rule and not the exception. What then was the "policy attraction" (Phillips 2004)? Why did OBE resonate in a system that was already saturated with regulations on teacher performance? Money drives educational reform in Mongolia. National programs and action plans are sometimes buried because of lack of external funding, and vice versa, new reforms are imported because of the promise of international funding. With that in mind, OBE must be regarded as a reform that, quite literally, was lucrative for all stakeholders involved. The teachers were given an additional bonus system to boost their low salaries, and government officials were credited for pumping money into the public sector.

Chapter 8 presents voucher-based (non-) reform as a case of discursive or fake borrowing where only the language of reform, but not the reform itself, was imported. Originally, the idea of vouchers held a great appeal because it signaled a new orientation toward choice and diversity in teacher in-service courses. In practice, however, they were merely used as (unequally distributed) registration forms. The phenomenon of fake policy borrowing is so common in Mongolia that it deserves to be treated in a larger political and economic context. As with more than 30 other postsocialist countries that became orphaned in 1990, government officials in Mongolia have learned to speak the "language of the new allies" (Silova 2004) and to profess the new values of market orientation necessary for adoption in the new international space. The chapter also analyzes a common reform strategy of international nongovernmental organizations (NGOs): the establishment of a parallel system of education that initially is fully financed by international donors, and is eventually supposed to become mainstreamed, institutionalized, and funded by government sources. Although the government-funded system selectively borrowed trainers, methods, and material used in the NGO-funded system, the wholesale adoption or handover did not materialize for a variety of reasons, including the voucher non-reform.

Chapter 9, finally, addresses a rare but vexing example of the encounter between local and global forces. We explore a non-encounter or a nonevent, and reflect on the following question: What if there is nothing to borrow? What if the transnational regimes have not built a portfolio of best practices, are not able to rely on an already tested reform package, and only know what they do not want to support, while lacking ideas of what to support? The example of nomadic education in Mongolia is an intriguing case for a situation of stalemate, in which the transnational regimes could not make up their minds. As a result, little was done to reerect or reform the dilapidating boarding-school system. We interpret the long decade of neglect (1991–2003) in nomadic education as a rare case of a transfer vacuum: the transnational regimes refused to fund a well-established system from the socialist past, and at the same time had no alternatives to offer. The tragedy of disregard is evidenced by the soaring dropout rates in rural areas. In studying dropouts one is astounded by how official educational statistics are tampered with to suit political and economic purposes. A section of chapter 9 therefore traces the "statistical eradication" of the dropout issue over the course of the past 15 years. Manufacturing educational statistics for political and economic reasons is not exclusively a Mongolian problem, but it has vast repercussions in a country that is under constant pressure to justify the need for external funding.[7] The fact that dropouts and non-enrollees almost disappeared from official statistics suggests that the government and international donor organizations decided not to treat these children with priority any longer. As opposed to the other case studies that scrutinize local reactions to global reforms, chapter 9 stresses global reactions to local practices. The study of nomadic education has urged us, perhaps more forcefully than our other case studies, to come to grips with the culture of international cooperation. What international donors make of local practices is as important as what meaning local actors attach to imported global reforms.

The "Global Speak" of Politicians and Its Impact on Local Practices

In recent years, educational transfer researchers have noticed that policy makers increasingly generate reform pressure by making reference to globalization, and resorting to broad terms like "international standards" in education. The boom in international student achievement studies (OECD and IEA studies) reflects the preoccupation with international benchmarks and the political interest in international comparison. Panic-stricken, no educational system wants to be "left behind." Whether globalization in education is real or imagined, it is uncontested that the "semantics of globalization" (Schriewer 2000: 330) is increasingly enlisted to accelerate educational reform. The intersection of globalization and borrowing/lending research has generated two new research questions, both of which we have treated with priority in this book: First, is it feasible to map the trajectory of educational transfer in an era of globalization? Second, how does educational import impact already existing policies?

First, the difficulty of mapping the trajectories of transplanted reforms has been highlighted by several scholars in globalization studies, and has rendered the spatial dimension of borrowing and lending research problematic. For example, is outcomes-based education (OBE), *originally* a New Zealand, Australian, Canadian, or U.S. reform? The answer varies, depending on the time period one is referring to, and on the expert one is asking. In the end, how valid is the genealogical approach to the study of reform epidemics that, as in the case of OBE, spread like wildfire around the globe? Today, OBE is as much a Chilean, South African, European, or Mongolian reform as it is a New Zealand or Australian reform. Late adopters of a reform, as is the case with the Mongolian government and over 30 other postsocialist governments, do not necessarily resort to the original(s), but rather orient themselves toward early adopters of the reform from their own world-system or "educational space" (Nóvoa and Lawn 2002). In contrast to nineteenth century borrowing research, when scholars could content themselves with tracing transplanted policies across the Atlantic (between North America and Europe), many scholars nowadays suggest giving up the traditional practice of actually mapping the itinerary of a traveling policy. Once we acknowledge multiple references for policy borrowing, we should perhaps move on to abandoning the mapping exercises in borrowing and lending research.

To make things more complicated, there is a special type of reference made in different parts of the world in recent years that has caught our attention. Politicians and policy makers increasingly make de-territorialized references to an imagined international community. As mentioned before, they generate reform pressure domestically by invoking fears of "falling behind" and urge their constituents to comply with "international standards" in education. What these international standards consist of has remained unclear. Nevertheless, these invocations are very effective in low-income countries, propelling a flurry of eclectic policy borrowing from wherever funds are made available. One approach to resolving the lack of explicit references is to acknowledge the existence of a "referential web" (Vavrus 2004), another is to recognize the blurred trajectories and label transplanted educational reforms simply as "traveling policies" (Seddon 2005; Coulby, Ozga, Popkewitz et al. 2006). In that sense, all global educational reforms qualify as traveling policies: one does not know where they come from, and go to; they are at the same time nobody's and everybody's reform. All one witnesses is that they surface at different times in different corners of this world.

Second, the question as to how existing practices are impacted by policy import is often brushed off with a general comment on hybridization. For example, the case studies in Anderson-Levitt's edited volume (2003), as remarkable as they are, focus exclusively on how a global reform such as outcomes-based education takes on a different meaning in various contexts.

The destiny of existing policies in light of such global forces is not explicitly addressed. Arguably, it is no small feat to examine how one and the same reform is reinterpreted differently as this tells us something about culture, and in particular about the culture of reform in the various policy contexts. However, hybridization resulting from the encounter between imported and already existing policies is but one of several conceivable outcomes. Other conceivable outcomes are a replacement of previous policies, and at the other extreme, a reinforcement of what had already been in place. Again, hybridization has been amply documented (e.g., Anderson-Levitt 2003), and replacement as an outcome of borrowing has also been well examined for societies that either underwent revolutionary (e.g., Spreen 2004) or other political changes (e.g., Luschei 2004). The first two groups of researchers who examine hybridization and replacement view globalization as a form of external intervention that inevitably triggers change. We may qualify this assumption by adding that, for a variety of political and economic reasons, so-called external interventions are frequently internally induced as politicians and policy makers utilize the semantics of globalization to generate reform pressure. But we are still left with those cases where policy import served exclusively to reinforce existing policies. Iveta Silova's study of bilingual education policies in post-Soviet Latvia (Silova 2005) is the only study to date that provides a solid empirical foundation to suggest that policy borrowing is sometimes used to legitimize and reinforce existing practices. Under which conditions policy import hybridizes, replaces, or reinforces existing practices is a key issue for globalization researchers, and has therefore been placed at center stage in our analyses.

The Lucidity of Fuzzy Predictions in Case Study Research

Arguably, our study on globalization in educational reform encapsulates only our observations in Mongolia. However, this book is a case study, and as such it mirrors the shortcomings and strengths of case study methodology. A shortcoming for some, but a strength for others, case studies make "fuzzy predictions" of the type "x in y circumstances *may* result in z" (Bassey 2001: 6; see also Hammersley 2001; Pratt 2003). The uncertainty of predictions relates to the contextual information in which a case study must be nested. Even though we have, every now and then, included observations made in other contexts, the bulk of our findings are deeply rooted in the Mongolian context of educational reform. This type of fuzzy prediction is, of course, different from statistical generalizations ("x in y circumstances results in z") used in quantitative comparative education research, and based on large and representative sample sizes (N).

The fuzziness of case studies is both a methodological weakness and strength. What accounts for the fuzziness is the complexity of an issue (many variables) that unfolds when attention is paid to the different actors, agendas, units of analysis, and practices within a context. In our case study, for example, we consistently used three units of analysis, corresponding to three distinct policy levels (policy talk, action, implementation); the lists of actors, agendas, and practices, varied for each chapter, and would be too long to reiterate here. Following Charles Tilly's advice (Tilly 1997: 5), we have constructed "causal stories" by investigating how the various actors, agendas, policy levels, and educational practices relate to each other. As a corollary, our case study must be regarded as a thick description based on a small N, but many variables (Ragin 1997).

This method of inquiry represents quite a different enterprise than what convergence theorists tend to undertake. Without exception, neoinstitutionalist theory rests upon a method that brings into focus long-term changes across many nations (e.g., Baker and LeTendre 2005; Boli and Thomas 1999). The method thrives upon a large N, and few variables. The tendency toward convergence that surfaces in such studies addresses a specific, precisely measurable layer of educational reform; in our opinion, at the expense of more complex layers. What the scholars of such large-scale studies encounter is an internationally shared understanding of modern schooling. This finding resonates with what we observed in our case study, and which we have chosen to label the "global speak" because it captures the universal language of reform spoken in Mongolia. As Francisco Ramirez, prominent comparativist and sociologist of neoinstitutional theory, astutely points out, "we are all in the business of sense-making" (Ramirez 2003: 240). What differentiates various approaches is the research focus. It is therefore more accurate to narrow the research focus by asking, "making sense of what?"

Convergence or world-culture theory acknowledges local variations of the global model of schooling, but regards them either merely as manifestations of loose coupling between official and enacted policy, or as part of a world-culture that encourages difference and diversity. It simply is not interesting to comparative sociologists to analyze how and why exactly the same school reform—let's say "vouchers"—is interpreted and implemented differently in various cultural contexts. There is little to gain for them in the way of making better sense of trends at system level. The fact that policy makers in different parts of the world justify choice, vouchers, privatization of education, and a host of other reforms in terms of "progress" and "justice," only reconfirms the theory on the international convergence in education. As mentioned before, we too have observed the trend toward "global speak" in educational reform. Indeed, the tower of Babel is transformed into a bazaar where global labels for educational reforms are traded. For us, the existence of "global speak" did not mark the culminating point of our findings, but rather the point of departure to dig deeper into why this universal language is spoken in Mongolia.

In this book, we have used the methodological tools of case study research to understand what "globalization" means to the various actors in Mongolia: why "it" is appropriated or rejected, and how "it" is adapted and modified to their cultural contexts. In the context of Mongolia (and other low-income countries), we tried to make sense of the phenomenon that traveling reforms are only borrowed rhetorically or very selectively. Understanding the political and economic gains that are associated with speaking a universal language of educational reform helped us assemble the puzzle of rhetorical or selective policy borrowing.

Analyzing in detail local policy contexts is paramount for understanding what globalization means for and does to people. We have tried to be specific in the usage of terms such as "local," "global," "internal," or "external." Among many other reasons, the practice of international organizations residing in a given country, or government officials making frequent study visits abroad, has rendered the use of these terms suspect. In concert with other researchers, we find spatial determinism of limited value (e.g., Camaroff and Camaroff 2001). Yet, "globalization" is frequently used as a justification to move educational reforms in a certain direction. This book has attempted to scratch at the surface of "global speak," and dig deeper into the politics and economics of an emerging lingua franca in educational reform. Not only in Mongolia but also in other countries that heavily depend on external funding from their new political allies, we have observed the contours of reform bilingualism: a universal language addressed to international donors, and a native language of reform that resonates with citizens. In a few cases of educational import in Mongolia, "global speak" functioned as a "killer language" (Skutnabb-Kangas 2000) displacing local variants of educational practices. In most cases, however, both languages are spoken simultaneously, one language for policy talk, and an another when it comes to selectively enacting and implementing imported policies.

Notes

Notes to Pages 17–28

1 Going Global: Studying Late Adopters of Traveling Reforms

1. In 2004, MFOS was dissolved and replaced with several Mongolian nongovernmental organizations. The educational programs of MFOS are since then carried out by the Mongolian Education Alliance and the Open Society Forum.
2. Our study on school-related migration in Dornod (Steiner-Khamsi, Stolpe, and Tümendelger, 2003), e.g., drew attention to the financial difficulties of small rural schools. We criticized the block grant finance system whereby schools receive funding based on their student numbers. The system has put rural schools, most of which are small with regard to student size, at risk, and they can barely survive and retain teachers and students. Starting in 2004, educational authorities in Mongolia acknowledge the difficulties of small schools and consider the size of school as a factor when they reallocate funds (Bat-Erdene 2005).

2 Educational Import in Mongolia: A Historical Perspective

1. Diamond Vehicle (Vajrayana) is a Tibetan form of Mahayana Buddhism.
2. The Yuan Dynasty refers to the period of Mongolian rule in China from the conquest of Chinggis Khan until 1368, i.e., until the end of the Mongolian Empire.
3. It should be noted here that the terms "Red Hat" and "Yellow Hat" are dated terms that give very imprecise descriptions of the different schools of Tibetan Buddhism. For the present context, however, they will suffice given that details of the internal relationships in Tibetan religious history cannot be addressed here.
4. His dates are sometimes also given as 1534–83.
5. "*Dalai*" is the Mongolian word for "ocean" or "sea." "Lama" (Mongolian: *lam*) derives from the Tibetan "*bla-ma*" and originally referred to a high- ranking religious teacher and dignitary who has the authority to pass on the teachings. The title "Dalai Lama" can be translated as "Priest of Immeasurable Wisdom" and has been the title of the reincarnated leader of the Lamaist church up until today, based on the original Mongolian-Tibetan hybrid form introduced by Altan Khan.
6. The number of monasteries founded gradually grew after proselytization. While only 5 monasteries existed in the Mongolian areas between 1601 and 1650, by 1750 there were an additional 69 new monasteries and in the following 100 years there were 114 more newly founded (Barkmann 2000: 51). By the beginning of the twentieth century there were over 700 monasteries in Mongolia.
7. According to Rinchen (1964: 28), at the end of the nineteenth century ca. 44% of the male population attended monastic schools.

8. The word *nom* means "book" or "knowledge" or "education" (often used in compound phrases: *erdem nom*) as well as "Buddhist teaching" or "canon" or "sutra" (often in combination: *burkhny nom*).
9. This territory, traditionally settled by Manchurians, stretches northeast of Mongolia from the Yenisei River to the Pacific and from Kamchatka to the Korean Peninsula.
10. Outer Mongolia represents approximately the area of today's Mongolia. According to Baabar (1999: 62), the terms "Inner" and "Outer" Mongolia were brought into use by the rulers of the Qing Dynasty to designate the territories within and without the areas under their direct control. From the perspective of Beijing, the Mongolian areas south of the Gobi are Inner and north of the Gobi Outer Mongolia.
11. "Suzerainty" is a term used by historians and political scientists to denote the supremacy over a semi-sovereign dependent state that is not in a position to act autonomously.
12. An administrative unit within the aimags introduced by the Manchurians. In Outer Mongolia where, according to baabar (1996: 93, 117), there was a total of 125 *khoshuu*, 1 *khoshuu* consisted of an average of 1,200 households and 6,000 inhabitants. The division into banners was abandoned in 1931.
13. Name of the Mongolian capital from 1706 to 1911 (also: Da Khüree).
14. Jigmedsüren and Baljirgarmaa (1966) have documented these decrees in a collection of sources, complemented with relevant newspaper articles.
15. Name of the Mongolian capital from 1911 to 1924.
16. For example, the Foreign Ministry invited the Englishman Uiding as a teacher in 1914 and sent out a letter with specifications for his arrival, accommodations, and pay. In the following year he was awarded a state medal together with his Buriat colleagues (including a female!) (Jigmedsüren and Baljirgarmaa 1966).
17. According to Shagdarsüren (1976: 27) and Baasanjav (1999: 326), in 1919 there were 60 secular schools in the country.
18. For further examples of the Buriat influence, see Moses and Halkovic (1985: 293ff.), Morozova (2002: 40ff., 70ff.), Bulag (1998: 12), and Bulag (2002: 150ff.).
19. The Comintern developed out of the First and Second Internationals founded at the end of the nineteenth century. These organizations were concerned with global problems such as war, colonialism, and questions of nationality. The Communist International was founded in 1919 at Lenin's initiative; it sought to establish a common political line within the communist parties via Marxist theory.
20. Classical Mongolian: *Sudur bichig-ün küriyeleng*.
21. This was also true for almost all other socialist countries.
22. Starting in 1939 textbooks were imported from Kazakhstan for the Kazakhs (Schöne 1973).
23. *Ardyn Ikh Khural*: Parliament of the MPR.
24. See chapter 3 for a detailed account of the "bypassing capitalism" narrative.
25. In the German Democratic Republic the polytechnic principle was even eponymous for comprehensive schools, which were called "Allgemeinbildende Polytechnische Oberschulen" (POS; "General Education Polytechnic High Schools").
26. More on this is found in chapter 6.
27. The most important primary and secondary teacher education newspapers were *Ardyn Gegeerel*, *Bagsh*, *Khüükhdiin Khümüüjil*, *Dund Surguul*, and *Surgan Khümüüjüülegch*.

3   Bypassing Capitalism

1. In an interview, Amgalan made some very personal remarks about the origin of the image. The white horse, he said, is the grown version of the foal in his 1951 work titled "Spring." He created the foal at the beginning of his artistic career, and thus he matured parallel to it (Enkhbold 1985). This parallel between the maturation of man and animal is a very common idea in the philosophy of Mongolian education. It is part of a very old tradition to give small children young animals that will grow to become the core of the child's own herd.
2. Kazakhstan, Kyrgyzstan, Tajikistan, Turkmenistan, Uzbekistan.
3. The Comintern followed the First and Second Internationals established at the end of the nineteenth century. These associations were concerned with the global situation, including questions of war, colonialism, and nationality. Lenin initiated the founding of the Communist International in 1919, within which the communist parties sought to develop a common political line in accordance with Marxist theory.
4. Analogous attempts aimed at deculturation via the ideologically motivated censorship of textbooks also occurred in Inner Mongolia. There, until the end of the 1970s, historically important Mongolian figures were either excluded from history lessons or, as in the case of Gada Meilin (also written as Gadameiren), their biographies were reduced to motifs of class struggle (Bao 1997: 212, 216; Bulag 2002: 143–149).
5. It should be noted that the approach in Mongolian historiography of labeling phases according to analogue criteria for democracy still persists in the reevaluation and working through of Mongolian history after 1990. An example of this appears in the most comprehensive historical volume of the post-1990 period, published in 1999 and coauthored by some of the most eminent Mongolian historians. The period from 1921 to the mid-1930s is characterized as a "phase in which the results of the national democratic revolution were expanded." But, in contrast to the socialist paradigm of interpretation, the next phase that lasted into the 1950s was called "the period of the totalitarian regime" (Mongol Ulsyn Ikh Surguul', Öchir Töv 1999: 24–25).
6. See chapter 2.
7. In postsocialist countries, during the phase in which new standards of measurement were being introduced in all social areas, it was common to start a new calendar with the year 1990, as an ironic answer to the proclaimed "end of history."
8. The children's drawings that appear in the HDR were created during a poverty awareness campaign led by the Mongolia Poverty Alleviation Program Office. Unfortunately, they were published anonymously.

4   Exchanging Allies: From Internationalist to International Cooperation

1. CMEA: Council for Mutual Economic Assistance, also known as Comecon: Council for Mutual Economic Cooperation. It was established in 1949 with the USSR, Bulgaria, Hungary, Poland, Romania, and Czechoslovakia as founding members. Albania entered later that year, but starting in 1961 it remained only as a formal member. The German Democratic Republic joined as a full member in 1950. Yugoslavia served as an observer starting in 1955 and became an associated member in 1964. The Mongolian People's Republic joined in 1962, followed by Cuba (1972) and Vietnam (1978), all of which became full members.

2. The preparations for the Complex Program for economic integration began in 1962 and the program was accepted by all members in 1971 (Rathmann and Vietze 1978: 347).
3. The first international pioneer camps were established in 1955. The most famous camp was the "Artek" on the Crimean Peninsula in the USSR. The pioneer republic "Wilhelm Pieck" in the GDR had international meetings with children's organizations including those from nonsocialist countries. A local attraction there was a yurt palace given as a gift by the Mongolian people. In the MPR, international pioneer meetings were held in the Nairamdal camp.
4. At the initiative of the Railroad Association's "Collective of Socialist Work," e.g., money was raised in the MPR for the National Children's Fund during work sessions (Tsedenbal-Filatova 1981: 105).
5. The Mongolian pioneer organization was named after the revolutionary hero Sükhbaatar.
6. Gorbachev's program of *perestroika* also had a positive impact on domestic policies and foreign relations in Mongolia. In 1987, Mongolia established formal relations with the United States, and in the same year signed agreements with the People's Republic of China settling border disputes. About one-quarter of the Soviet army stationed in Mongolia was withdrawn in 1987, and the government of the USSR planned to recall the entire contingent from Mongolia by 1990. Morris Rossabi (2005) presents a detailed account of the *perestroika* period in Mongolia (1987–1990) during which opposition leaders were able to freely express their criticism of the government.
7. Starting in the 1970s, the Mongolian government periodically requested from the Soviet government an annulment of their previous debts and loans (Enkhtor 2004). Between 1996 and 2000 the Democratic Union refused to acknowledge any financial obligations toward the Russian Federation that, in turn, prompted the Russian government to collect $120 million in debt interest per year (Borchuluun 2004). In January 2004, the MPRP-led government finally resolved the so-called Big Debt in diplomatic negotiations with the Russian government, and had the debt cancelled by making a controversial transfer of $200 million to a secretive bank account in Russia (Bolormaa 2004).
8. But it is of course a concern for countries that are heavily indebted due to recurring loans from international financial institutions. The 1996 World Bank initiative on "heavily indebted poor countries" (HIPC) provides information on the 38 HIPC countries (the majority of African countries, and a few countries in Central and Latin America, and in Southeast Asia), and on conditions of debt relief (World Bank 2004b).
9. Ulziisaikan (2004: 2) introduces an interesting comparison: The annual interest for government loans is only 1%, whereas individuals in business are charged with an annual interest rate of 36–42% from commercial banks. In addition, commercial banks require the debtor to act as the "keyplayer for the whole duration of the loan," and require a variety of documents, warrants, and forms of collateral. This begs the question of why ordinary business people are subject to tight scrutiny and accountability, while the government is allowed not only a lower annual interest rate, but also less public accountability?
10. A third education sector development project is in preparation. In December 2004, the government and ADB signed a contract for a technical assistance grant (total: $625,000). The consultants, hired from this fund, will prepare the content and cost of the project. The grant is mostly financed by the Japan Special Fund, and administered by ADB (ADB 2004a).

11. The total project cost of $68.5 million is funded as follows: $45 million Government of Japan, $14 million ADB, $4.8 million Nordic Development Fund, and $4.7 Government of Mongolia (ADB 2002: iii). 95% of the ESDP is financed by loans, and only 5% by a technical assistance grant.
12. Although we see the pedagogical value of an exercise in which decision makers learn about the dire needs of rural schools, we object to using schools as a human zoo.

5 Structural Adjustment Reforms, Ten Years Later

1. Similarly, the IMF also engaged in a comprehensive review of its loan conditionality in 2000–2002. The outcome was a series of new guidelines on conditionality (IMF 2002) that emphasize, among other things, ownership and capacity to implement programs in the countries that borrow from the IMF. Koeberle, Silarsky, and Verheyen (2005) discuss the tensions between conditionality and national ownership from an interesting historical perspective.
2. In the second half of the 1990s, poverty remained constant at 36% of the population despite the large Poverty Alleviation Program (Government of Mongolia, World Bank, and UNDP 1999). At the same time, income inequality rose significantly between 1995 and 2002 (Government of Mongolia and UNDP 2004: 27).
3. Note that John Weidman (2001) does not share our opinion on the unilateral approach of ADB. In our definition, SWAp is among others characterized by multiple funding sources; a feature that the first two education loans of ADB were lacking. For example, the financing plan for the Second Education Sector Program was as follows (Government of Mongolia and ADB 2001: 51): ADB finances 76% of the Second Education Sector Program (95% loan, 5% technical assistance grant), and the rest is funded by the government.
4. Economists use different methods to assess the reduction of external assistance as a percentage of the GDP in the early 1990s. See, e.g., the (much lower) figures provided by the Government of Mongolia and UNDP (2000).
5. Apart from the Open Society Institute (Soros Foundation) that deals with the postsocialist region, the international nongovernmental organizations operating in the Mongolian education sector also erroneously assign Mongolia to the East Asian and Pacific region. Save the Children UK, e.g., includes Mongolia in the list of 19 countries in the Southeast, East Asia, and Pacific region.
6. Albania, Kazakhstan, and Slovakia are not included in Figure 5.1 due to incomplete time series data.
7. Until 1992 educational spending as a percentage of all public expenditures was over 25%, and then it fell in 1993 to 15.6%. Since then, government spending in education has continuously increased, and since 2000 the percentage of public expenditure spent on education has been constant at around 20%.
8. Figure from 2000 (Steiner-Khamsi and Nguyen 2001: 9).
9. One of the common critiques, e.g., is that the current educational finance systems forces schools to "heat out the window" (Bartlett, Byambatsogt, and Enkhamgalan 2004: 8) because heating costs are part of the fixed and guaranteed portion of the budget, whereas maintenance cost (including fixing broken windows) are part of the variable cost that include numerous other posts such as salaries (at least 70% of the variable budget), and social insurances (15%). The inefficiency critique addresses two issues: First, schools are constantly forced to submit requests for a budget amendment, because their submitted budget was drastically reduced. Second, re-allocations of the approved budgets are a

common practice undoing all previously made agreements and commitment. The Public Expenditure Tracking Survey (PETS) of the World Bank is examining some of these critiques in more detail.
10. The figures for the unofficial economy range from 15 (Government of Mongolia and UNDP 2004) to 32.5 as a percentage of the official GDP (Amar 2004).
11. Officially, every member of the *negdel* (animal husbandry collective) of 16 years of age and older was permitted to privately own 10 head of cattle. The maximum per household was 50. These restrictions varied for the different vegetation zones. In the Gobi area each member was allowed to possess 15 head of cattle, and the household was confined to a maximum of 75 (Bawden 1968: 402; Barthel 1990: 117).
12. According to the Regulation 42 of February 18, 2004 (Mongol Ulsyn Zasgiin Gazar 2004), the lowest monthly base salary for teachers is T 47,000, and the highest is T 59,400. The figure T 639,400 is the average of an annual base salary of a regular teacher.
13. In 2003, the government spent T 6,7 billion ($5.9 million) on student scholarships. 57.4% of the recipients were children of civil servants, 25.1% children from poor families, 8.9% children from herders. The remaining 8.6% of all government scholarships were distributed to disabled or orphaned children, or gifted students (Otgonjargal 2004: 28). The children of civil servants, including children of teachers, receive the bulk of government scholarships.

6 The Mongolization of Student-Centered Learning

1. The basic information provided in all opening meetings is as follows: the year the school was built, the number of school facilities, the number of school staff (distinguished between administrative staff, teachers, and support staff), the number of students, the number of students in the boarding school, the number of shifts, the percentage of herder families, the sources of heat and energy, and the awards received by the school, the staff, and the students. Some schools also volunteer information on their "quality rate" (number of students that passed the grade 4 and grade 8 standardized exams with a grade A or B), their "success rate" (number of students that passed these exams), and their ranking at provincial level.
2. The inverse also applies, and was in fact very pronounced in daily life under socialism: to make fun of political or public events in private.
3. There are several explanations for the status loss of the class oldest in recent years. One of the most convincing is that today the oldest student in the classroom is probably there for reasons related to poverty. They enrolled late or had to interrupt their studies because their parents could either not afford to send them to school, or they were needed to help with herding.
4. In our data analysis we compared (former) class monitors that served before the 1990s with current class monitors. The historical perspective enabled us to understand how the roles of class monitors were affected by political changes. Before 1990 there was a division of labor between class monitors and the communist organization of youth pioneers. The former was focused on the classroom and the school, whereas pioneers were mainly active during after-school events. After the abolishment of the pioneer organization in schools, the class monitors were overwhelmed from having to take on the extra task of organizing after-school events. This book chapter does not elaborate on the historical dimension of the study.

5. In one school in Bayan-Ölgii we interviewed the total population of class monitors (all class monitors from grade 1 to 10), but in other educational institutions (including in teacher education institutions) we drew convenience samples, i.e., interviewed all class monitors that were available on the day of data collection.
6. Unlike postsocialist Mongolia, where the class monitor system evokes overall positive sentiments, educators in other postsocialist countries are not as enthusiastic about the system. In one of our informal discussions, colleagues from Hungary, Slovakia, and Croatia shared their experiences in school where the class monitors were in cahoots with the pioneer organization. As they remembered it, class monitors frequently denounced their peers, and functioned as mind-police.
7. The unemployment rate for women far exceeds the rate for men (National Statistical Office of Mongolia 2002: 53ff.), and the impact on female-headed households is alarming (UNICEF 2000). Nevertheless, there is much more discussion of male unemployment and the public perception of male unemployment deserves closer scrutiny. It could be interpreted either as gender discrimination (insensitivity toward women's right to economic independence) or as a stigmatization of unemployed men, often associated with alcoholism and vagrancy.
8. The exact figures for the overrepresentation of female teaching staff are as follows (ADB and World Bank 2004: 47): 93.6% in primary, 71.1% in lower secondary, 68.5% in upper secondary schools, and 60.1% in vocational and technical schools. The overrepresentation decreases with each level of education, and in colleges and universities only 52.4% of all lecturers are female.
9. The study was a follow-up study of TIMSS in which researchers recorded and analyzed mathematics and science lessons in Germany, Japan, and the United States.
10. The syllabus for the course Comparative Study of Educational Systems, taught by James E. Russell at Teachers College, Columbia University, serves as a good illustration. Russell taught the course for the first time in academic year 1899–1900, and made students in his class compare "characteristic features" of one educational system with that of another (cited in Bereday 1963: 189).

7 Outcomes-Based Education: Banking on Policy Import

1. We are referring here to two of the studies that included, among other research questions, teachers' perception of OBE in Mongolian schools: the Teachers as Parents study, and the pilot study for the Public Expenditure Tracking Survey (see chapter 1).
2. OBE was discussed in the United States for a brief period in the late 1980s, but was replaced by Goals 2000 in most states, a standards-based approach to curriculum reform issued by the federal government in 1994. However, the 2001 No Child Left Behind Act (NCLB) triggered a renewed interest in outcomes-based accountability or "new accountability," a term coined by researchers affiliated with the Consortium for Policy Research in Education (Fuhrman 1999). See Jennifer O'Day's critique of the "outcomes-based bureaucratic model of school accountability" (O'Day 2002: 294), which she finds reemerging in federal legislation in general, and the NCLB Act in particular.
3. John Smyth and Alastair Dow (1998: 291) label OBE the "new educational orthodoxy" of the 1990s when educational outcomes were propagated as the only effective regulative mechanism over schooling.

4. Its notoriety is captured in the Swiss German expression *helvetische Verspätung* (Helvetic/Swiss procrastination), commonly used with a trace of self-irony.
5. All along, however, the public administration reform in Mongolia was accompanied by numerous warnings. For example, the ADB review of governance in Mongolia comments on the outcomes-contracts as follows (ADB 2004b: 17): "To be implemented effectively, the contract-based Public Sector Management and Finance Law requires a strong, rules-based government that enforces contracts and is characterized by robust markets—conditions that for the most part do not yet exist in Mongolia. . . . It would be foolhardy to entrust public managers with complete freedom over resources when they have not yet internalized the habit of spending public money according to prescribed rules."
6. The government "saved" T 7 billion (approximately $6.2 million) from this exercise, and schools were at a loss as to how to balance the payments made for the month of January.
7. Article 47 deals with "Assessment of Performance Agreement" and Article 49 with "Payment of Performance Bonuses to Employees" (Parliament of Mongolia 2002).
8. The greatest delay occurred at the parliamentary level. The first draft for the Public Sector Management and Finance Act was submitted in 1997 (see Lanking 2004), but it was only approved in 2002. Once it was approved the ministries were eager to adopt it with the financial support of ADB to their sector.
9. The monthly base salary ranges from T 53,200 ($47) for a regular teacher to T 63,840 ($57) for a teacher at the rank of an advisor (Mongol Ulsyn Zasgiin Gazar 2004: 274; see also Shinjlekh Ukhaan, Bolovsrolyn, Khün Amyn Bodlogo, Khödölnöriin, Sangiin Said 1995: 220–224).
10. Today the social worker takes part in school administration and is, among other things, in charge of coordinating extracurricular activities, and monitoring teachers' attitudes and behavior in school. During the socialist era these roles were assigned to the leader of the communist pioneer organization in the school.
11. In one school in Ulaanbaatar: drunkenness calls for a 20–40% salary deduction, depending on how often the teacher has shown up drunk in school in the past month.
12. The teacher holds a monopoly on the usage of red ink.
13. The soaring private cost of education is a topic that frequently makes headlines in the media. Apart from official private costs such as tuition in higher education, textbook, and stationary fees, and the purchase of school uniforms, there is a host of educational expenses that are unofficial but yet tolerated by school administrators, ranging from small parental "donations" (e.g., being pressured to buy gifts for a teacher's birthday, or purchase booklets that teachers, school administrators, or methodologists produced) to sizeable expenses (e.g., private tutoring for exam preparation, or bribery for getting admitted to university entrance exams).
14. At the beginning of the school year, the teacher must formulate academic benchmarks for each student. The rubric "student development" only requires a statistical summary of grade fluctuation in the class. OBE critics in Mongolia have made the argument that the establishment of academic benchmarks at the beginning of the school year encourages grade inflation at the end.
15. In her study of 278 teachers in Ethiopia, Fenot Aklog (2005) found widespread criticism toward data-driven accountability and teacher career ladders. The ladder was introduced in 1998, and resembles the Mongolian teacher salary scheme in great detail. In Ethiopia, the performance evaluations are equally viewed as corrupt and disruptive to the collegial atmosphere in the schools.

16. Drawing from a random sample of over 100 schools in Mongolia, PETS gathered (among other data) information on financial redistribution or reallocation practices at the school, provincial, and central levels. At the school level, the World Bank study examined salary supplements and deductions, as well as bonuses and bonus reductions. The first results of PETS will be released in 2006.

8 Speaking the Language of the New Allies with the Voucher (Non-) Reform

1. The category "administrators" includes *aimag* governors, directors of the social policy units, and directors of the Education and Culture Centers at the provincial level. The visits and interviews covered schools in the following provinces: Arkhangai, Bayankhongor, Töv, Dundgov', Ömnögov', Khovd, Bayan-Ölgii, and Dornod.
2. N. Enkhtuya (from 1998 to 2004 Mongolian Foundation for Open Society, since 2004 Mongolian Education Alliance), was instrumental in codesigning and co-analyzing the data from the surveys, and A. Gherelmaa, Open Society Forum, Ulaanbaatar, helped with the translation of the policy documents.
3. The statistical material on in-service training is presented in the annual evaluation reports of the School 2001 project, available at the Mongolian Education Alliance.
4. The national programs are referred to as *ündesnii khötölbör*, and guidelines are labeled either *chiglel* or *khögjüülekh khötölbör*.
5. Changes in this high turnover rate are expected. The outgoing MPRP-led government insisted that no replacements are made based on party affiliation after the elections in summer 2004. The new government, a thin majority of representatives from a broad and unstable Motherland-Democracy coalition, accepted the deal but created new layers of administration to secure influence in the new administration.
6. Sneath's study was conducted in cooperation with the Sociological Research Center of the Mongolian Institute of Administration and Management Development.
7. *Shovgor* means "conical" or "tapered," and the term, according to Sneath, emerged in the socialist period, and was used as a euphemism for offering vodka by referring to the shape of the bottle (Sneath 2002b: 87).
8. Citation from the article "*Emch bagsh khoyort avilga ög. Yaadag yum be?*" ("Give Doctors and Teachers Bribes! Why Not?") published in *Ödriin Sonin*, December 19, 2001, and quoted in Sneath (2002b: 96).
9. As part of the USAID-funded Central Asian project Participation, Education, and Knowledge Strengthening, commonly referred to as PEAKS.

9 What if There is Nothing to Borrow? The Long Decade of Neglect in Nomadic Education

1. *Otor* indicates a temporary dwelling that is far away from the rest of the nomadic household. Usually several male members of the nomadic household move the herd to distant, more nutritious pastures, and establish, for a limited period, these types of satellite camps.

2. For a brief period in the early and mid-1990s, there was an inverse migration movement from cities to rural areas. Some of these "new nomads" became herders out of necessity because they faced unemployment in Ulaanbaatar or in the province towns, and at the same time lacked social networks that could supply them with food or cash. Others engaged in animal husbandry to improve their income.
3. Between 1993 and 2004 there were a total of six natural disasters (very cold and long winters, followed or preceded by a drought in the summer) (Grayson and Munkhsoyol 2004: 47). Millions of animals perished during these natural disasters *(zud)* forcing herder families, especially those with a small livestock, to leave the rural areas and seek an alternative income in semi-urban and urban areas.
4. Mongolia is the world leader for mutton and goat meat supply per capita (FAO 2004). In 2002, the annual mutton and goat meat supply per capita was 144 pounds in Mongolia, followed by Iceland, as a distant second, with 55 pounds, and New Zealand with 54 pounds placed third. In comparison, the per capita mutton and goat meat supply in the United States is ranked 264. Most likely, the consumption figures are slightly lower than the supply figures because of wastage.
5. Among other distinctions, Mongolian herders also distinguish between animals that graze far away from the camp as opposed to those that stay nearby. Horses and camels are considered "long leg animals," because of their wide radius of roaming. In contrast, cattle, sheep, and goats are referred to as "short leg animals" because they graze near the camp.
6. The calculation for 1 head of livestock is based on sheep *(bog)*. The Public Perception Survey (Grayson and Munkhsoyol 2004) counted 1 camel or 1 horse as 7 sheep (or 7 heads of livestock), and 1 cow as 6 sheep. In the Mongolian literature it is actually more common to use a horse or a cow as a unit of measurement *(bod)*, and determine the heads of livestock according to that unit. In this text, however, we use the *bog* measurement used in the public perception survey on nomadic pastoralism (Grayson and Munkhsoyol 2004) in order to present their findings.
7. The National Board for Children, Save the Children UK, and UNICEF commissioned the study *The living conditions of the children in peri-urban areas of Ulaanbaatar* (National Board for Children, Save the Children UK, and UNICEF 2003) that focused on poor children in Ulaanbaatar. In contrast to dropouts from rural areas (del Rosario, Battsetseg, Bayartsetseg et al. 2005), the children in Ulaanbaatar were not ashamed but angry and disappointed that they were told not to go to school any more. The lack of interest or, "doesn't like to learn" explanation for dropouts rarely came up in the Ulaanbaatar study (National Board for Children, Save the Children UK and UNICEF 2003: 27f.) or in the rural study on dropouts (del Rosario, Battsetseg, Bayartsetseg et al. 2005).
8. There exist no exact statistics on children and youth that are physically or mentally impaired. The Ministry of Education (MOSTEC, UNDP, UNESCO et al. 2000: 45) points out that in 1997, 8% or 34,000 of all school-aged children were physically or mentally impaired: 37% of these children were not enrolled in schools, 5.8% were placed in special schools (there exist only 3 schools for children with special needs nationwide), and the remaining half (17,000) attended regular schools without any special provision.
9. This is a comment on how the necessity for a National Poverty Alleviation Program was justified, and not on the fact that such a program was initiated in the first place. It is important to bear in mind that the commitment of the government to poverty reduction is one of the conditionalities for receiving loans from international financial institutions (World Bank, Asian Development Bank, and more recently, International Monetary Fund).

10. For more than a decade DANIDA was the only organization committed to preserving and improving schools in remote rural areas. Beginning in 1992, DANIDA supported rural school development. Their recurring grants, however, were not primarily directed toward the rehabilitation of boarding schools, but rather focused on improving the overall learning environment and the quality of education in rural schools.

10   Bending and Borrowing in Mongolia, and Beyond

1. DANIDA, e.g., is not as concerned with visibility, and in fact, several DANIDA-funded projects in Mongolia are "wrongly" attributed to UNESCO (Gobi Women Initiative) or the government (nonformal education program). In contrast MFOS, a newcomer in 1997, was concerned with visibility and manageability (availability of "local capacity" to implement projects), but less with cost-effectiveness.
2. For more details, see chapter 5.
3. The belief in a period of "socialist transformation" or "socialist transition" also prevailed in other socialist countries, and was in some cases differentiated by region. In the People's Republic of China, e.g., Central China embarked on the collectivization campaign during the First Five-Year Plan (1953–1957), whereas Xinjian and Inner Mongolia only completed the socialist reconstruction by 1959. Due to the rebellions against the Chinese government in Tibet, Mao postponed the "democratic reform" in Tibet to the Third Five-Year Plan (1963–1967), and collectivization in the Tibet Autonomous Region (TAR) began a decade later than in China. The "transition" period in Central China lasted, according to Chinese historical accounts, 4 years, in Xinjian and Inner Mongolia 10 years, and in TAR 14 years (Bass 1998: 29f.).
4. The Asian Development Bank has given the largest amount of loans to the Government of Mongolia, and yet ADB's commitment to the social sector (education, health, social services) only accounts for 9% of all their financial activities in Mongolia (ADB 2004a).
5. With the reorganization of the Soros Foundation Network in Mongolia in 2004, the following two Mongolian nongovernmental organizations became the successor NGOs of the educational programs, previously based at the Mongolian Foundation for Open Society: Mongolian Education Alliance (MEA), and Open Society Forum (OSF).
6. CMEA, also known as Comecon, was established in 1949 and dissolved in 1991 (see chapter 4). In the same year in 1949, Harry Truman appealed in his inaugural address as president of the United States to "advanced nations" to support the economic development in "underdeveloped areas" of the world. The concept of a "fair deal" whereby richer nations help poorer nations by means of technology and knowledge transfer is often referred to as the Truman doctrine (cited in Escobar 1995: 3).
7. LeTendre, Akiba, Goesling et al. (2000) use the term "policy trap" to denote the tendency of government officials to generate reform pressure by exaggerating the findings from international comparative studies. Their study focuses on how the findings from the Third International Mathematics and Science Study (TIMSS) were interpreted by U.S. government offices.

References

Abu-Lughod, J. L. 1989. *Before European hegemony: The world system, A.D. 1250–1350.* New York: Oxford University Press.

ADB [Asian Development Bank]. 2000. *Country assistance plan (2001–2003)—Mongolia.* Manila, Philippines: ADB.

———. 2002. *Proposed loan Second Education Development Project (Mongolia)*, dated July 16. Manila, Philippines: Board of Directors, Asian Development Bank.

———. 2003. *ADB approves loans for second phase of public sector reforms in Mongolia*, dated October 14, 2003. Available at: <http://www.adb.org/Documents/News/2003>. [Accessed January 22, 2005].

———. 2004a. *Technical assistance to Mongolia for preparing the third education development project.* Grant Number TAR: MON 34187. Manila, Philippines: Asian Development Bank.

———. 2004b. *Governance: Progress and challenges in Mongolia.* Manila: Asian Development Bank.

———. 2005. *East and Central Asia region.* Available at: <http://adb.org/EastCentralAsia/default.asp>. [Accessed February 10, 2005].

ADB [Asian Development Bank] and World Bank. 2004. *Country gender assessment Mongolia.* Draft report March 2004 (author of draft: Helen T. Thomas). Manila and Washington: ADB and World Bank.

Agvaan, C., and C. Bat. 1981. Mongolia and the developing countries. In *The 60th Anniversary of People's Mongolia*, 149–152. [Name of editor not noted]. Ulaanbaatar: Unen Editorial Board, and Moscow: Novosti Press Agency Publishing House.

Aklog, F. 2005. *Sources of teacher job satisfaction and dissatisfaction: A study of urban primary school teachers in Ethiopia.* Ed.D. dissertation. New York: Teachers College, Columbia University.

Alexijewitsch, S. 1999. *Seht mal, wie ihr lebt. Russische Schicksale nach dem Umbruch.* [Look how you are living. Russian fates after the radical changes]. Berlin: Aufbau Taschenbuch Verlag. (Original title in Russian: Zaocharovannye smert"yu [Under the spell of death]).

Amar, N. 2004. *Shadow economies: Size, growth and consequences.* Ulaanbaatar: Bank of Mongolia.

Amarkhüü, O., ed. 1968. *Ardyn bolovsrol, soyol urlag, shinjlekh ukhaany talaarkhi khuul' togtoomjiig sistemchilsen emkhtgel. 1964–1967 on, III.* [Systematic compilation of laws and decrees in education, culture and art in the period 1964–1967, volume 3]. Ulaanbaatar. [Publisher not noted].

Anderson, J. 1998. *The size, origins, and character of Mongolia's informal sector during the transition.* World Bank policy research working paper number 1916. Washington, DC: World Bank.

Anderson, J. D. 1988. *The education of blacks in the South, 1860–1935.* Chapel Hill and London: University of North Carolina Press.

Anderson-Levitt, K. 2003. A world culture of schooling? In *Local meanings, global schooling. Anthropology and world culture theory*, edited by K. Anderson-Levitt, 1–26. New York: Palgrave Macmillan.
Andreassjan, R., and A. Eljanow. 1968. Probleme der Strukturgestaltung und Industrialisierung in den Entwicklungsländern [Problems of structure building and industrialization in developing countries]. *Sowjetwissenschaft. Gesellschaftswissenschaftliche Beiträge* (7): 738–750.
Appadu, K., and N. Frederic. 2003. *Selected readings on sectorwide approaches (SWAp) and education*. Paris: UNESCO International Institute for Educational Planning.
Appadurai, A. 1990. Disjuncture and difference in the global cultural economy. In *Global culture*, edited by M. Featherstone, 295–310. London: Sage Publications.
———. 1997. Globale Landschaften [Global landscapes]. In *Perspektiven der Weltgesellschaft*, edited by Ulrich Beck, 46–67. Frankfurt: Suhrkamp.
———. 2000. Grassroots globalization and the research imagination. *Public Culture* 12 (1): 1–19.
Ardyn Bolovsrolyn Yaam [Ministry of People's Education]. 1976. *Bügd Nairamdakh Mongol Ard Ulsyn Ardyn Bolovsrolyn Khögjilt. Statistik emkhtgel.* [Development of the people's education in the Mongolian People's Republic. Compilation of statistics]. Ulaanbaatar: Ardyn Bolovsrolyn Yaamny khevlel.
Ariunaa, Sh. 2003. *The new stage of development of NGOs.* Available at: <http://www.opensocietyforum.mn>. [Accessed March 17, 2005].
Atwood, C. 2004. *Is Mongolia part of Central Eurasia?* Presentation at the Conference of the Mongolia Society, October 16. Bloomington: University of Indiana.
Baabar (Bat-Erdeniin Batbayar). 1996. *XX Zuuny Mongol. Nüüdel Suudal. Garz olz* [Mongolia in the 20th century. Removal and settling. Losing and getting]. Ulaanbaatar. [Publisher not noted].
———. 1999. *History of Mongolia*. Cambridge, UK: White Horse Press.
Baasanjav, Z. 1999. Surgan khümüüjüülekh setgelgee, surguul', bolovsrol. [Educational thought, school, and education]. In *Mongol ulsyn tüükh*, 321–326. Ulaanbaatar: Mongol Ulsyn Ikh Surguul', Öchir töv. [Editor not noted].
Baker, D. P., and G. K. LeTendre. 2005. *National differences, global similarities. World culture and the future of schooling*. Stanford: Stanford University Press.
Bale, M., and T. Dale. 1998. Public sector reform in New Zealand and its relevance to developing countries. *The World Bank Research Observer* 13: 103–121.
Bao, W. 1997. *When is a Mongol? The process of learning in Inner Mongolia*. Dissertation. University of Washington and Michigan: UMI Dissertation Service, Bell & Howell Company.
Barkey, K., and M. von Hagen, eds. 1997. *After empire: Multiethnic societies and nation-building*. Boulder: Westview.
Barkmann, U. B. 1999. *Geschichte der Mongolei oder Die "Mongolische Frage." Die Mongolen auf ihrem Weg zum eigenen Nationalstaat*. [The history of Mongolia, or the "Mongolian question." The Mongols on the path to their own nation-state]. Bonn: Bouvier.
———. 2000. *Landnutzung und historische Rahmenbedingungen in der Äusseren Mongolei / Mongolischen Volksrepublik (1691–1940)*. [Land usage and historical conditions in Outer Mongolia / The Mongolian People's Republic (1691–1940)]. Osaka: National Museum of Ethnology.
Barthel, H. 1990. *Mongolei—Land zwischen Taiga und Wüste* [Mongolia—country between taiga and desert]. Geographische Bausteine, Neue Reihe, Heft 8. Gotha: VEB Hermann Haack Geographisch-Kartographische Anstalt.

Bartlett, W., Mr. Byambatsogt, and Mr. Enkh-amgalan. 2004. *Technical assistance for improved education expenditure in Mongolia. Final report.* Washington, DC: World Bank.
Bass, C. 1998. *Education in Tibet. Policy and practice since 1950.* London: Zed Books.
Bassey, M. 2001. A solution to the problem of generalisation in educational research: Fuzzy predictions. *Oxford Review of Education* 27 (1): 5–22.
Bataa, D. 1998. *Belcheeriin mal aj akhuig erkhelj irsen mongolchuudyn ulamjlalt arga.* [Mongolian traditional methods of pastoral animal husbandry]. Ulaanbaatar. [Publisher not noted].
Batbayar, Ts. 2003. Foreign policy and domestic reform in Mongolia. *Central Asian Survey* 22 (1): 45–49.
Batdelger, J., G. Dulamjav, D. Enkhtuya, S. Enkhtuvshin, D. Indra, D. Kesler, E. Narmandakh, N. Norjkhorloo, and Sh. Orosoo. 2000. *Evaluation of Mongolia's National Programme for Preschool Strengthening (NPPS), 1995–2000.* Ulaanbaatar: Ministry of Science, Technology, Education and Culture.
Bat-Erdene, R. 2005. *Baseline study of Mongolia's educational budget preparation and execution stages.* Ulaanbaatar: Open Society Forum and World Bank.
Battogtokh, D. 2002. *Mongol surguuliin tovchoon. Mongol ardyn geriin surguul'. Tergüün devter.* [Compendium on Mongolian schools. Yurt schools of the Mongolian people, volume 1]. Ulaanbaatar: BSShUYa-ny khar"ya Bolovsrolyn Khüreelen, Alban Bus Bolovsrol Töv.
Baumert, J., E. Klieme, M. Neubrand, M. Prenzel, U. Schiefele, W. Schneider, P. Stanat, K.-J. Tillmann, and M. Weiss, eds. 2001. *PISA 2000: Basiskompetenzen von deutschen Schülerinnen und Schülern im internationalen Vergleich.* [PISA 2000: Basic competencies of German students in international comparison]. Opladen: Leske & Budrich.
Bawden, C. R. 1968. *The modern history of Mongolia.* London: Weidenfeld and Nicholson.
Bayangol. 2004a. *Bayangol düürgiin surgalt üildverleliin "Setgemj" tsogtsolboryn angi udirdsan bagshiin nemegdel khölsiig tootsoj olgokh juram.* [Regulation on renumeration and eligibility of salary supplements for class teachers at the "Setgemj" complex school in Bayangol district]. Ulaanbaatar: Bayangol city-district, Setgemj complex school.
———. 2004b. *Bayangol düürgiin "Setgemj" tsogtsolbor surguuliin bagsh naryn devter zasaltyg khyanaj, ünelekh juram.* [Regulation on how to control and evaluate the notebook correction of teachers at the "Setgemj" complex school in Bayangol district]. Ulaanbaatar: Bayangol city-district, Setgemj complex school.
Bayasgalan, S. 1990. Surguuliin zarim asuudlyn tukhai. [About a few problems of schools]. *Khödölmör* August 23: 3.
Bazargür, D., B. Chinbat, and S. Shiirev-Ad"yaa. 1989. *Bügd Nairamdakh Mongol Ard Ulsyn malchdyn nüüdel.* [Pastoral nomadism in the Mongolian People's Republic]. Ulaanbaatar: Ulsyn Khevleliin Gazar.
———. 1992. *Territorial organisation of Mongolian pastoral livestock husbandry in the transition to a market economy.* PALD Research Reports 1. Brighton: University of Sussex, IDS.
Beck, L. 2005. World Bank social accountability project in Mongolia. Personal communication, dated April 20, 2005. New York.
Begz, N. 2002. Globalchlalyn üyeiin Mongol ulsyn bolovsrolyn khögjliin onol, argazüin ündsen asuudluud. [Theoretical and methodological issues regarding the development of public education in Mongolia in the era of globalization]. *Bolovsrolyn Sudlal* 1: 24–29.

Bentaouet Kattan, R., and N. Burnett. 2004. *User fees in primary education.* Education for All working papers. Washington, DC: World Bank.
Bereday, G. Z. F. 1963. James Russell's syllabus of the first academic course in comparative education. *Comparative Education Review* 7 (2): 189–196.
Berg-Schlosser, D. 2002. Comparative studies: method and design. *International encyclopedia of the social and behavioural sciences,* 2427–2433, no. 4. Amsterdam: Elsevier.
Bikeles, B., Ch. Khurelbaatar, and K. Schelzig. 2000. *The Mongolian informal sector: Survey results and analysis.* Ulaanbaatar: Economic Policy Support Project, USAID.
Boli, J., and G. M. Thomas. 1999. INGOs and the organization of world culture. In *Constructing world culture: International non-governmental organizations since 1875,* edited by J. Boli and G. M. Thomas, 13–49. Stanford: Stanford University Press.
Bolormaa, L. 2004. *Secret bow tie celebration following Big Debt.* Available at: <http://www.opensocietyforum.mn>. [Accessed March 14, 2005].
Bolormaa, Ts. 2001. Determinants and facilitating factors of migration: Retrospective reports from migrants to Ulaanbaatar, Darkhan-Uul and Tuv Aimag. In *A micro study of internal migration in Mongolia,* edited by the National University of Mongolia, 20–29. Ulaanbaatar: National University of Mongolia, Population Teaching and Research Center.
Boone, P. 1994. Grassroots macroeconomic reform in Mongolia. *Journal of Comparative Economics* 18: 329–356.
Borchuluun, Ya. 2004. *Secret bow tie celebration following Big Debt.* Available at: <http://www.opensocietyforum.mn>. [Accessed March 14, 2005].
Bormann, K. D. 1982. Zu einigen Aspekten der politisch-staatlichen Entwicklung der MVR bei der Errichtung des Sozialismus. [Several aspects of the political-state development of the MPR during the establishment of socialism]. In *Die Mongolische Volksrepublik. Historischer Wandel in Zentralasien,* edited by author collective, 76–93. Berlin: Dietz-Verlag.
Brook Napier, D. 2003. Transformation in South Africa: Policies and practices from ministry to classroom. In *Local meanings, global schooling,* edited by K. Anderson-Levitt, 51–74. New York: Palgrave Macmillan.
Bruun, O., and O. Odgaard. 1996. A society and economy in transition. In *Mongolia in transition. Old patterns, new challenges,* edited by O. Bruun and O. Odgaard, 23–41. Nordic Institute of Asian Studies, Studies in Asian Topics, no. 22, Richmond: Curzon Press.
Bruun, O., P. Ronnas, and L. Narangoa. 1999. *Country analysis Mongolia. Transition from the Second to the Third World?* Stockholm: Swedish International Development Cooperation Agency, Asia Department.
Bulag, U. E. 1998. *Nationalism and hybridity in Mongolia.* Oxford: Clarendon Press.
———. 2002. *The Mongols at China's edge. History and politics of national unity.* Lanham, Boulder, New York, Oxford: Rowman & Littlefield.
Camaroff, J., and J. L. Camaroff. 2001. Millennial capitalism: First thoughts on a second coming. In *Millennial capitalism and the culture of neoliberalism,* edited by J. Camaroff and J. L. Camaroff, 1–56. Durham, NC: Duke University Press.
Carnoy, M., and P. McEwan. 2001. Privatization through vouchers in developing countries: The cases of Chile and Colombia. In *Privatizing education: Can the market place deliver choice, efficiency, equity, and social cohesion?* edited by Henry M. Levin, 151–177. Boulder, CO: Westview.
Carnoy, M., and D. Rhoten. 2002. What does globalization mean to educational change? A comparative approach. *Comparative Education Review* 46 (1): 1–9.

Carnoy, M., R. Jacobsen, L. Mishel, and R. Rothstein. 2005. *The charter school dust-up.* Washington, DC: Economic Policy Institute, and New York: Teachers College Press.
Chabbott, C. 2003. *Constructing education for development: International organizations and Education for All.* New York: RoutledgeFalmer.
Chagdaa, Kh. S. 2004. *Mongol ger büliin khariltsaany ulamjlal shinechlel.* [Tradition and change in Mongolian family relations]. Ulaanbaatar: Mönkhiin üseg.
Changai, B. 1974. Schülerselbstverwaltung und allseitige Entwicklung der Persönlichkeit des Schülers. [Student self-governance and holistic development of student personality]. In *II. Konferenz der Pädagogen sozialistischer Länder*, edited by Akademie der Pädagogischen Wissenschaften der Deutschen Demokratischen Republik, 110–116. Berlin: Akademie der Pädagogischen Wissenschaften der DDR..
Chase-Dunn, C., and T. D. Hall. 1997. *Rise and demise: Comparing world systems.* Boulder, CO: Westview.
Child, B. J. 1998. *Boarding school seasons. American Indian families 1900–1940.* Lincoln and London: University of Nebraska Press.
Chimeddorj, M. 1997. *Report on the activities of Education and Science Trade Unions of Mongolia. Report of the president.* Ulaanbaatar: Mongolian Education and Science Trade Union.
Chisholm, L. 2005. The politics of curriculum review and revision in South Africa in regional context. *Compare* 35 (1): 79–100.
Choimaa, Sh., L. Terbish, D. Bürnee, and L. Chuluunbaatar. 1999. *Buddyn shashin, soyolyn tailbar tol'.* [Annotated dictionary of Buddhist religion and culture]. Buddyn sudlal tsuvral II, negdügeer devter. Ulaanbaatar. [Publisher not noted].
Chubb, J. E., and T. M. Moe. 1992. *A lesson in school reform from Great Britain.* Washington, DC: Brookings Institution.
Chuluunbaatar, L. 2002. *Nüüdelchin mongolchuudyn bichig üsgiin soyol.* [The script culture of nomadic Mongols]. Ulaanbaatar: MUIS-iin Buddyn Soyol Sudalgaany Töv, London dakh' Tövdiin Fond.
CICE [Current Issues in Comparative Education]. 1998. Are NGOs overrated? Special issue of *Current Issues in Comparative Education* 1(1), online journal. Available at: <http://www.tc.columbia.edu/cice>. [Accessed November 24, 2005].
———. 2001. Sector-wide approaches in education: Coordination or chaos? Special issue of *Current Issues in Comparative Education* 3(2), online journal. Available at: <http://www.tc.columbia.edu/cice>. [Accessed November 24, 2005].
Clifford, J., and G. E. Marcus. 1986. *Writing culture. The poetics and politics of ethnography.* Berkeley and Los Angeles: University of California Press.
Confederation of Mongolian Trade Unions. 1995. *Letter to Mr. Jose Ayala-Lasso, United Nations High Commissioner for Human Rights*, dated July 3, 1995. Ulaanbaatar: Confederation of Mongolian Trade Unions.
Connell-Szasz, M. 1979. *Education and the American Indian. The road to self-determination since 1928.* Albuquerque, NM: University of New Mexico Press (first published in 1974).
Coulby, D., J. Ozga, T. S. Popkewitz, and T. Seddon, eds. 2006. *World yearbook in education 2006: Education research and policy.* London and New York: Routledge.
Cowen, R. 1999. Late modernity and the rules of chaos: An initial note on transitologies and rims. In *Learning from comparing. New directions in comparative educational research*, edited by R. Alexander, P. Broadfoot, and D. Phillips, 73–88. Oxford: Symposium.
Cuban, L. 1998. How schools change reforms. *Teachers College Record* 99 (3): 453–477.

Cummings, W. K. 2003. *The institutions of education. A comparative study of educational development in the six core nations.* Oxford: Symposium Books.
Dale, R. 2001. Constructing a long spoon for comparative education: Charting the career of the "New Zealand Model." *Comparative Education* 37: 493–500.
Damdinsüren, Ts. 1959. *Soyolyn öviig khamgaal"ya.* [Let's preserve our cultural heritage]. Ulaanbaatar. [Publisher not noted].
Dashdavaa, Ch. 1999. Medleg ukhaan, shinjlekh ukhaany khögjil. [The development of knowledge and science]. In *Mongol ulsyn tüükh,* 309–321. Ulaanbaatar: Mongol Ulsyn Ikh Surguul', Öchir töv. [Editor not noted].
Dashtseden, T. 1984. *BNMAU dakh' mergejliin bolovsrol.* [Vocational education in the MPR]. Ulaanbaatar: BNMAU, Ardyn bolovsrolyn yaamny surakh bichig, setgüüliin negdsen redaktsiin gazar.
Dashzeveg, B. 1971. BNMAU-d khüree, khiid, lam naryn asuudlyg shiidverlesen baidal. [How the question of monasteries and lamas was resolved in the MPR]. In *BNMAU-yn Kapitalist bus khögjliin tüükhen turshlagyn zarim asuudal,* 200–214. Ulaanbaatar. [Editor and publisher not noted].
del Rosario, M., D. Battsetseg, B. Bayartsetseg, Ts. Bolormaa, S. Dorjnamjiin, S. Tumendelger, and Ts. Tsentsenbileg. 2005. *The Mongolian dropout study.* Ulaanbaatar: Mongolian Education Alliance.
Denemark, R. A., J. Friedman, B. K. Gills, and G. Modelski, eds. 2000. *World System history. The social science of long-term change.* London and New York: Routledge.
Department of Science and Education of the Samara Region. 2001. *Ob izmenenii stoimosti imennogo obrazovatel'nogo cheka No. 372-od ot 29. 12. 2001.* [(Russian) Regulations on the voucher system in teacher in-service training. Appendix to decree number 372 of December 29, 2001]. Samara, Russian Federation: Department of Science and Education.
Desai, P. 1997. Introduction. In *Going global. Transition from plan to market in the world economy,* edited by P. Desai, 1–94. Cambridge, MA: MIT Press.
———. 2002. *Financial crisis, contagion, and containment. From Asia to Argentina.* Princeton: Princeton University Press.
Dimou, A. 2004. Alter Wein in neuen Flaschen? Darstellungen von Sozialismus in deutschen Schulbüchern. [Old wine in new bottles? Portrayals of socialism in German textbooks]. *Internationale Schulbuchforschung* 26: 347–363.
Dondog, T. 1974. Über grundlegende Massnahmen der Mongolischen Revolutionären Volkspartei zur Persönlichkeitsbildung in der sozialistischen Gesellschaft während der letzten Jahre. [The fundamental measures of the Mongolian People's Revolutionary Party towards personality development in socialist society in the last few years]. In *II. Konferenz der Pädagogen sozialistischer Länder. Arbeitsprotokoll—Diskussion der Sektionen—Teil I,* 131–133. Berlin: Akademie der Pädagogischen Wissenschaften der DDR.
Donnelly, K. 2002. *A review of New Zealand's school curriculum. An international perspective.* Wellington, New Zealand: Education Forum.
Dorzhsuren, Y. 1981. The Flowering of Art and Culture. In *The 60th Anniversary of People's Mongolia,* 108–122. [Name of editor not noted]. Ulaanbaatar: Unen Editorial Board, and Moscow: Novosti Press Agency Publishing House.
Dugersuren, M. 1981. An equal and active member of the United Nations. In *The 60th Anniversary of the People's Republic of Mongolia,* 141–146. [Name of editor not noted]. Ulaanbaatar: Unen Editorial Board, and Moscow: Novosti Press Agency Publishing House.

Dugger, C. W. 2004. In Africa, free schools feed a different hunger. *New York Times*, October 24.
Dyer, C. 2001. Nomads and Education for All: Education for development or domestication? *Comparative Education* 37 (3): 315–327.
Eade, D. 2003. *Capacity-building. An approach to people-centered development.* Oxford: Oxfam UK and Ireland (first published in 1997).
Edwards, M., and D. Hulme. 1996. Too close for comfort? The impact of official aid on nongovernmental organizations. *World Development* 24 (6): 961–973.
Eggert, K. 1970. 50 Jahre Volksbildungswesen in der Mongolischen Volksrepublik. Eindrücke einer Studienreise. [50 years people's educational system in the Mongolian People's Republic. Impressions from a study visit]. *Vergleichende Pädagogik* (6): 413–418.
Egelund, N. 2002. The causes of school dropouts in Mongolia. In *School development in Mongolia 1992–2000*, edited by E. Nørgaard, 69–90. Copenhagen, Denmark: Copenhagen International Centre for Educational Development.
Elias, N. 1987. *Involvement and detachment.* New York: Blackwell.
Enkhbold, Ts. 1985. D. Amgalantai khiisen yariltslaga. [Interview with D. Amgalan]. *Khödölmör* 56: 4.
Enkhtor, G. 2004. *While Government celebrates, some politicians doubt Russian debt has retreated.* Interview with P. Jasrai, former Prime Minister of Mongolia. Available at <http://www.opensocietyforum.mn>. [Accessed on March 14, 2005].
Erdene-Ochir, G. 1991. *Khüneer khün khiikh. Mongol ukhaany survalj.* [Creating human beings with the help of human beings. The origin of Mongolian sciences]. Ulaanbaatar: Bolovsrolyn Yaamny surakh bichig, khüükhdiin nom khevleliin gazar.
Erüül Mendiin Yaam [Ministry of Health] and UNDP [United Nations Development Programme]. 2000. *Mongolyn ösvör üyeiinkhnii kheregtseeg sudalsan sudalgaany tailan.* [Adolescent needs assessment report]. Ulaanbaatar: Erüül Mendiin Yaam and UNDP, Admon Press.
Escobar, A. 1995. *Encountering development: The making and unmaking of the Third World.* Princeton, NJ: Princeton University Press.
FAO [Food and Agricultural Organization]. 2004. *Food balance sheet, year 2002.* Available at: <http://www.faostat.org>. [Accessed on May 2, 2005].
Fernández-Giménez, M. E. 1997. *Landscapes, livestock, and livelihoods: Social, ecological and land use change among the nomadic pastoralists of Mongolia.* Ph.D. dissertation. Berkeley, CA: University of California-Berkeley.
———. 2001. The effects of livestock privatization on pastoral land use and land tenure in post-socialist Mongolia. *Nomadic Peoples* 5 (2): 49–66.
Finke, P. 2000. *Changing property rights systems in Western Mongolia.* Max Planck Institute for Social Anthropology Working Papers, no. 3. Halle/Saale, Germany: Max Planck Institute.
Fisher, W. F. 1997. Doing good: The politics and antipolitics of NGO practices. *Annual Review of Anthropology* 26: 439–464.
Foner, E. 2002. *Who owns history? Rethinking the past in a changing world.* New York: Hill and Wang.
Forman, W., and B. Rintschen. 1967. *Lamaistische Tanzmasken. Der Erlik-Tsam in der Mongolei.* [Lamaist dance masks. The Erlik-Tsam in Mongolia]. Leipzig: Koehler & Amelang.
Foucault, M. 1984. What is an author? In *The Foucault reader*, edited by P. Rabinow, 101–120. New York: Pantheon.

Foucault, M. 1995. *Discipline and punish: The birth of the prison.* New York: Vintage (2nd edition).
Frank, A. G. 1992. *The centrality of Central Asia.* Amsterdam: University of Amsterdam, Centre for Asian Studies.
Fratkin, E. 1997. Pastoralism: Governance and development issues. *Annual Review of Anthropology* 26: 235–261.
Freud, S. 1965. *Jokes and their relation to the unconscious.* New York: Norton (first published 1905).
Fuhrman, S. H. 1999. *The new accountability.* Philadelphia: Consortium on Policy Research in Education.
Fukuyama, F. 1993. *The end of history and the last man.* New York: Free Press.
Galsan, D. 1981. The Sühe Baator Young Pioneer Organisation. In *The 60th Anniversary of People's Mongolia*, 126–128. [Name of editor not noted]. Ulaanbaatar: Unen Editorial Board, and Moscow: Novosti Press Agency Publishing House.
Gataullina, L. M. 1981. *Stroitel'stvo socialisticheskoi kultury v Mongolskoi Narodnoi Respublike.* [(Russian) The establishment of socialist culture in the Mongolian People's Republic]. Moskva: Izdatel'stvo "Nauka" glavnaya redaktsiya vostochnoi literatury.
Giordano, C., and D. Kostova. 2002. The social production of mistrust. In *Postsocialism. Ideals, ideologies and practices in Eurasia*, edited by C. M. Hann, 74–91. London and New York: Routledge.
Gladwell, M. 2002. *The tipping point.* Boston: Little and Brown.
Goldstein, D. M. 2003. *Laughter out of place. Race, class, violence, and sexuality in a Rio shantytown.* Berkeley and Los Angeles: University of California Press.
Government of Mongolia. 1993. *Mongolia sector review. Mongolia human resource development and education reform project,* dated December 21, 1993. Ulaanbaatar: Ministry of Science and Education.
———. 2001a. *Interim poverty reduction strategy paper.* Ulaanbaatar: Government of Mongolia.
———. 2001b. *Project and technical assistance proposals, 2001–2004.* Paris, France, May 15–16: Mongolia consultative group meeting. Ulaanbaatar: Government of Mongolia.
Government of Mongolia and ADB [Asian Development Bank]. 2001. *Second Education Development Project. Final report,* dated February 23. Stockholm, Sweden: SWEDEC International AB.
———. 2002. *Loan agreement for the Second Education Development Project,* dated August 16. Ulaanbaatar: ADB.
Government of Mongolia and UNDP [United Nations Development Programme]. 1997. *Human development report Mongolia 1997.* Ulaanbaatar: Free Press Printing House.
———. 2000. *Human development report Mongolia 2000.* Ulaanbaatar: UNDP.
———. 2004. *Human development report Mongolia 2004.* Ulaanbaatar: UNDP.
Government of Mongolia, World Bank, and UNDP [United Nations Development Programme]. 1999. *Mongolia independent evaluation of the National Poverty Alleviation Program and options post-2000.* Report by A. Batkin, Ms. Bumhorol, R. Mearns, and J. Swift. Washington, DC: World Bank.
Grayson, R., and B. Munkhsoyol. 2004. The future of nomadic pastoralism in Mongolia. Public perception survey. Ulaanbaatar: Open Society Forum.

Gundsambuu, Kh. Kh. 2002a. *Mongolyn niigmiin davkhraajil: khögjil, khandlaga.* [Social stratification in Mongolia: Developments and tendencies]. Ulaanbaatar: Mongol Ulsyn Zasgiin Gazryn Kheregjüülegch Agentlag Udirdlagyn Akademi.

———. 2002b. *Social stratification in contemporary Mongolian society.* Ulaanbaatar: "Zotol" Club of Professional Sociologists.

Günsen, R. 1962. *Bügd Nairamdakh Mongol Ard Uls ardyn ardchilsen sotsialist uls bolj khuvirsan n'.* [How the MPR was transformed into a people's democratic socialist state]. Ulaanbaatar: Ulsyn khevleliin khereg erkhlekh khoroo.

Hambly, G., ed. 1991. *Zentralasien. Weltgeschichte Band 16.* [Central Asia. World history, volume 16]. Frankfurt/M: Fischer Taschenbuch Verlag.

Hammersley, M. 2001. On Michael Bassey's concept of the fuzzy generalisation. *Oxford Review of Education* 27 (2): 219–225.

Hann, C. M., ed. 2002. *Postsocialism. Ideals, ideologies and practices in Eurasia.* London and New York: Routledge.

Hardt, M., and A. Negri. 2000. *Empire.* Cambridge, MA: Harvard University Press.

Harke, H., and M. Dischereit. 1976. *Geographische Aspekte der sozialistischen ökonomischen Integration.* [Geographical aspects of socialist economic integration]. Gotha/Leipzig: VEB Hermann Haack, Geographisch-Kartographische Anstalt.

Harley, K., and V. Wedekind. 2004. Political change, curriculum change and social formation, 1990 to 2002. In *Changing class. Education and social change in post-apartheid South Africa,* edited by L. Chisholm, 195–220. Cape Town, South Africa: HSRC Press, and New York: Palgrave Macmillan.

Henig, J. 1994. *Rethinking social choice: Limits of the market metaphor.* Princeton: Princeton University Press.

Heyneman, S. P. 2003. The history and problems in the making of education policy at the World Bank 1960–2000. *International Journal of Educational Development* 23: 315–337.

———. 2004. One step back, two steps forward. The first stage of the transition for education in Central Asia. In *The challenges of education in Central Asia,* edited by S. P. Heyneman, and A. J. DeYoung, 1–8. Greenwich, CT: Information Age Publishing.

Hobsbawm, E., and T. Ranger, eds. 1983. *The invention of tradition.* Cambridge, UK: Cambridge University Press.

Holmes, B. 1981. *Comparative education: Some considerations of method.* London: George Allen and Unwin.

Hopkin, A. G., ed. 2002. *Globalization in world history.* New York and London: W. W. Norton & Company.

Huff, D. 1954. *How to lie with statistics.* New York: W. W. Norton & Company.

Humphrey, C. 1978. Pastoral nomadism in Mongolia: The role of herdsmen's cooperatives in the national economy. *Development and Change* 9 (1): 133–160.

———. 2002a. Does the category "postsocialist" still make sense? In *Postsocialism. Ideals, ideologies and practices in Eurasia,* edited by C. M. Hann, 12–15. London and New York: Routledge.

———. 2002b. *The unmaking of Soviet life: Everyday economics after socialism.* Ithaca, NY: Cornell University Press.

Humphrey, C., and R. Mandel. 2002. The market in everyday life: Ethnographies of postsocialism. In *Markets and moralities: Ethnographies of postsocialism,* edited by R. Mandel and C. Humphrey, 1–16. Oxford, UK: Berg.

Humphrey, C., and D. Sneath. 1999. *The end of nomadism? Society, state and the environment in Inner Asia*. Durham: Duke University Press.
Idshinnorov, S., ed. 1997. *Jamsrany Tseveenii mendelsnii 115 jiliin oid. Tüüver zokchioluud, 1-r bot'*. [The 115th Anniversary of Jamsrany Tseveen. Collected works, volume 1]. Ulaanbaatar. [Publisher not noted].
Ilon, L. 1994. Structural adjustment and education: Adapting to a growing global market. *International Journal of Educational Development* 14 (2): 95–108.
IMF [International Monetary Fund]. 2002. *Guidelines on conditionality*. September 25, 2002. Washington, DC: IMF.
———. 2004. *IMF conditionality*. A factsheet—September 2004. Washington, DC: IMF.
Innes-Brown, M. 2001. Democracy, education, and reform in Mongolia: Transition to a new order. In *Education and political transition: Themes and experiences in East Asia*, edited by M. Bray and W.-O. Lee, 77–99. Hong Kong: Comparative Education Research Centre.
Ischi-Dordji. 1929. Die heutige Mongolei II: Kulturelle Aufbauarbeit in der Mongolei. [Contemporary Mongolia II: Establishing culture in Mongolia]. *Osteuropa. Zeitschrift für die gesamten Fragen des europäischen Ostens* 4 (6): 401–409.
Jagchid, S., and P. Hyer. 1979. *Mongolia's culture and society*. Boulder, CO: Westview Press, and England: Dawson Folkestone.
Jansen, J., and P. Christie, eds. 1999. *Changing curriculum: Studies on outcomes-based education in South Africa*. Kenwyn, South Africa: Juta.
Janzen, J., and D. Bazargur. 1999. Der Transformationsprozess im ländlichen Raum der Mongolei und dessen Auswirkungen auf das räumliche Verwirklichungsmuster der mobilen Tierhalter. Eine empirische Studie. [The transformation process in rural areas of Mongolia and its impact on spatial life patterns of mobile herders. An empirical study]. In *Räumliche Mobilität und Existenzsicherung. Fred Scholz zum 60. Geburtstag*, edited by J. Janzen, 47–82. Berlin: Institut für Geographische Wissenschaften, Freie Universität Berlin.
Jessipow, B. P., ed. 1971. *Pädagogik. Lehrbuch für Einrichtungen zur Ausbildung von Lehrern der unteren Klassen*. [Education. Textbook for teacher education institutions of lower school grades]. Berlin: Volk und Wissen Volkseigener Verlag.
Jigmedsüren, S., and B. Baljirgarmaa, eds. 1966. *Mongolyn avtonomit üyeiin surguul' (1911–1920)*. [The schools of Mongolia's autonomy period (1911–1920)]. Ulaanbaatar: Ulsyn khereg erkhlekh khoroo.
Jones, G. 1995. *How to lie with charts*. San Jose, CA: IUniverse Inc.
Jones, P. W. 1988. *International policies for Third World education: Unesco, literacy and development*. London and New York: Routledge.
———. 1998. Globalisation and internationalism: Democratic prospects for world education. *Comparative Education* 34 (2): 143–155.
———. 2004. Taking the credit: Financing and policy linkages in the education portfolio of the World Bank. In *The global politics of educational borrowing and lending*, edited by G. Steiner-Khamsi, 188–200. New York: Teachers College Press.
Jügder, Ch. 1987. *Mongold feodalizm togtokh üyeiin niigem–uls tör, gün ukhaany setgelgee*. [Public-state and philosophical ideas during the period of feudal stagnation in Mongolia]. Ulaanbaatar: Shinjlekh ukhaany akademi.
Kausylgazy, N. 1990. Surguuliin togtoltsoo yaaj öörchlökh ve? [How should the school system be transformed?]. *Ünen* January 18: 3.
Kazakevich, I. S. 1978. Die marxistisch-leninistische Theorie des nichtkapitalistischen Entwicklungsweges und ihre Realisierung. [The Marxist-Leninist theory of

non-capitalist development and its realization]. In *Der revolutionäre Weg der Mongolischen Volksrepublik zum Sozialismus. Probleme der Umgehung des kapitalistischen Entwicklungsstadiums*, edited by H.-P. Vietze, 175–212. Studien über Asien, Afrika und Lateinamerika, Band 31. Berlin: Akademie-Verlag.

Keuffer, J. 1991. *Buddhismus und Erziehung. Eine interkulturelle Studie zu Tibet aus erziehungswissenschaftlicher Sicht.* [Buddhism and education. An intercultural study of Tibet from an educational research perspective]. Münster/New York: Waxmann.

Khazanov, A. M. 1994. *Nomads and the outside world.* Madison: University of Wisconsin Press (first published in 1984).

Kim, B.-Y. 2003. Informal economy activities of Soviet households: Size and dynamics. *Journal of Comparative Economics* 31: 532–551.

Koeberle, S., P. Silarszky, and G. Verheyen. 2005. *Conditionality revisited: Concepts, experiences, and lessons learned.* Washington, DC: World Bank.

Korsun, G., and P. Murrell. 1995. Politics and economics of Mongolia's privatization program. *Asian Survey* 15 (5): 472–486.

Kotkin, S., and B. A. Elleman. 1999. *Mongolia in the twentieth century. Landlocked cosmopolitan.* Armonk: M. E. Sharpe.

Kratli, S. 2000. *Education provision to nomadic pastoralists. A literature review.* Washington, DC: The World Bank (also published in 2001 as working paper 126 of the Institute for Development Studies, Sussex).

Kunzmann, M. 1981. Zur Entwicklung der Berufsbildung in der Mongolischen Volksrepublik. [The development of vocational education in the Mongolian People's Republic]. *Berufsbildung* 11: 507–510.

Kuznetsov, D. V., and G. A. Kashoian. 1963. *Neoproverzhimoe.* [Russian Beyond all doubt]. Moscow: Molodaya gvardya.

Ladd, H. F., ed. 1996. *Holding schools accountable: Performance-based reform in education.* Washington, DC: Brookings Institution.

Ladd, H. F., and E. B. Fiske. 2003. Does competition improve teaching and learning? Evidence from New Zealand. *Educational Evaluation and Policy Analysis* 25 (1): 97–112.

Lanking, R. 2004. *Don't try this at home? A New Zealand approach to public management reform in Mongolia.* Manuscript. Wellington, New Zealand: Graduate School of Business and Government Management.

Lattimore, O. 1962. *Nomads and commissars. Mongolia revisited.* New York: Oxford University Press.

Lchamsüren, B. 1978. Die internationalistische Politik der MRVP. [The internationalist politics of the MPRP]. In *Der revolutionäre Weg der Mongolischen Volksrepublik zum Sozialismus. Probleme der Umgehung des kapitalistischen Entwicklungsstadiums*, edited by H.-P. Vietze, 359–380. Studien über Asien, Afrika und Lateinamerika, Band 31. Berlin: Akademie-Verlag.

Ledeneva, E. 1998. *Russia's economy of favours. "Blat," networking and informal exchange.* Cambridge: Cambridge University Press.

Lenhart, V. 1993. *"Bildung für alle." Zur Bildungskrise in der Dritten Welt.* [Education for All. The literacy crisis in the Third World]. Darmstadt: Wissenschaftliche Buchgesellschaft.

Lenin, V. I. 1945. *Chto dyelat'? Nabolevshiye voprosy nashego dvisheniya.* [(Russian) What to do? Burning questions of our movement]. Moskva: Politisdat.

LeTendre, G., M. Akiba, B. Goesling, A. Wiseman, and D. Baker. 2000. The policy trap: National educational policy and the Third International Math and Science

Study. *International Journal of Educational Policy, Research and Practice* 2 (1): 45–64.
Levin, B. 1998. An epidemic of education policy: (What) can we learn from each other? *Comparative Education* 34 (2): 131–141.
Lhagve, S. 1997. *Final report on "Internal markets in education": Think tank facility project MON/07/131*. Ulaanbaatar: Ministry of Finance of Mongolia and UNDP Field Office in Mongolia.
Lkhagvajav, Ch. 1997. Mongol ulsyn bolovsrolyn togtoltsoony shinechlel, khögjliin khandlaga. [Reforms and development trends in the Mongolian educational system]. In *Mongol ulsyn bolovsrol, Emchtgel No. 5*, edited by Mongol Ulsyn Gegeerliin Yaam, 20–21. Ulaanbaatar: Mongol Ulsyn Gegeerliin Yaam.
Lkhagvasüren, G., and J. Boldbaatar. 1999. Mongolchuudyn shashin shütleg. [The religiosity of Mongolians]. In *Mongol ulsyn tüükh*, 346–378. Ulaanbaatar: Mongol Ulsyn Ikh Surguul', Öchir töv.
Lindblad, S., and T. S. Popkewitz, eds. 2004. *Educational restructuring. International perspectives on traveling policies*. Greenwich: Information Age Publishing.
Lipski, V. 1963. Book review of *Nepoproverzhimoe*. [Beyond all doubt], written by D. V. Kuznetsov, and G. A. Kashoian. *Comparative Education Review* 7 (1): 95.
Luhmann, N. 1990. *Essays on self-reference*. New York: Columbia University Press.
Luschei, T. F. 2004. Timing is everything: The intersection of borrowing and lending in Brazil's adoption of "Escuela Nueva." In *The global politics of educational borrowing and lending*, edited by G. Steiner-Khamsi, 154–167. New York: Teachers College Press.
Luvsanchultem, N. 1981. For peace and friendship between nations. In *The 60th Anniversary of the People's Republic of Mongolia*, 25–140 [Name of editor not noted]. Ulaanbaatar: Unen Editorial Board, and Moscow: Novosti Press Agency Publishing House.
Lynch, J. 1998. The international transfer of dysfunctional paradigms. In *Learning and teaching in an international context: Research, theory and practice*, edited by D. Johnson, B. Smith, and M. Crossley, 7–33. Bristol: University of Bristol, Centre for International Studies in Education.
Mandel, R. 2002. Seeding civil society. In *Postsocialism. Ideals, ideologies and practices in Eurasia*, edited by C. M. Hann, 279–296. London and New York: Routledge.
Mandel, W. 1949. Outer Mongolia's Five-Year Plan. *Far Eastern Survey* 15 (June): 140–144.
Markow, L. 1968. Die RGW-Länder 1967. [The CMEA countries 1967]. *Sowjetwissenschaft. Gesellschaftswissenschaftliche Beiträge* (7): 751–755.
McEwan, P. 2000. The potential impact of large-scale voucher programs. *Review of Educational Research* 7 (2): 103–149.
McGinn, N. 1996. Education, democratization, globalization: Challenges for comparative education. *Comparative Education Review* 40 (4): 341–357.
Meyer, J. W., and F. O. Ramirez. 2000. The world institutionalization of education—origins and implications. In *Discourse formation in comparative education*, edited by J. Schriewer, 111–132. Frankfurt/M: Lang.
Ministry of Education [of New Zealand]. 1993. *The New Zealand curriculum framework*. Wellington, New Zealand.
———. 2001. *Curriculum stocktake report to Minister of Education*. Wellington, New Zealand: Ministry of Education. Available at: <http://www.minedu.govt.nz>. [Accessed on April 15, 2005].

MOECS [Ministry of Education, Culture and Science]. 2002. *Education Law*, May 3. Ulaanbaatar: MOECS.
———. 2003a. *A compilation of laws, resolutions, decrees and decisions related to the Education, Culture and Science Sector budget and finance*. Ulaanbaatar: MOECS Second Education Development Program.
———. 2003b. *Statistical overview in education, culture, arts, and technology*. Ulaanbaatar: MOECS.
———. 2005. *Educational statistics Mongolia, school years 1991/2–2004/05* (data base). Ulaanbaatar: MOECS, Department of Monitoring.
MOECS [Ministry of Education, Culture and Science] and ADB [Asian Development Bank]. 2001. *Education Sector Development Program*. Mongolian/English brochure. Ulaanbaatar: ADB Program Implementation Unit.
Mongol Messenger. 1997. [Name of author not noted]. Call for increased focus on education. *Mongol Messenger* August 20: 7.
Mongol Ulsyn Bolovsrol, Soyol, Shinjlekh Ukhaany Yaam [Mongolian Ministry of Education, Culture, and Science]. 2001. *Mongol ulsyn bolovsrolyn salbaryn üüsel, khögjil. 80 jild—Foundation of education sector in Mongolia and its development in 80 years. 80th Anniversary*. [Bilingual book]. Ulaanbaatar: Mongol Ulsyn Bolovsrol, Soyol, Shinjlekh Ukhaany Yaam—MOSTEC, Government building III.
Mongol Ulsyn Ikh Surguul', Öchir töv [Author collective]. 1999. *Mongol ulsyn tüükh*. [History of the state of Mongolia]. Ulaanbaatar. [Publisher not noted].
Mongol Ulsyn Zasgiin Gazar [Government of Mongolia] and UNDP [United Nations Development Programme]. 1997. *Mongolyn khünii khögjliin iltgel 1997*. [Human development report 1997]. Ulaanbaatar: Mongol Ulsyn Zasgiin Gazar and UNDP.
Mongol Ulsyn Zasgiin Gazar [Government of Mongolia]. *Mongol ulsyn zasgiin gazryn togtool 2004. Dugaar 42: Töriin alban khaagchiin tsalingiin talaar avakh arga khemjeenii tukhai*. [Decree of the Government of Mongolia, no. 42: Regulation on salaries of civil servants]; dated February 18. Ulaanbaatar: Government of Mongolia.
Mongolia Ministry of Science and Education and Academy for Educational Development. (1993–1994). *Human development and education reform project: Masterplan*. Ulaanbaatar: Ministry of Science and Education.
Mongolia National Government. 1998. *Joint decree of the Ministry of Finance and the Ministry of Education of Mongolia, number 62/125*, dated March 24. Ulaanbaatar: Mongolia National Government.
———. 2001. *National program for pre-service and in-service training of primary and secondary education teachers, appendix 1 to the government decree number 120*, dated June 5. Ulaanbaatar: Mongolia National Government.
Mongolia National Statistical Office. 2001. *2000 Population and housing census: The main results*. Ulaanbaatar: Government building III.
Mongolian People's Republic Council of Ministers and State Committee for Information, Radio and Television. 1978. Visit of friendship and brotherhood. *Mongolia* 5 (44): 5.
Monmonier, M. 1991. *How to lie with maps*. Chicago: University of Chicago Press.
Montgomery, Y.-Kh. D., and R. Montgomery. 1999. The Buriat alphabet of Agvan Dorzhiev. In *Mongolia in the twentieth century. Landlocked cosmopolitan*, edited by S. Kotkin, and B. A. Elleman, 79–97. Armonk, New York; London, England: M. E. Sharpe.
Morgan, D. 1986. *The Mongols*. Cambridge, MA: Basil Blackwell Ltd.

Morozova, I. Y. 2002. *The Comintern and revolution in Mongolia*. Cambridge: White Horse Press.
Moses, L. W. 1977. *The political role of Mongol Buddhism*. Bloomington: Indiana University.
Moses, L., and S. A. Halkovic Jr. 1985. *Introduction to Mongolian history and culture*. Indiana University Uralic and Altaic Series, volume 149, Research Institute for Inner Asian Studies, Bloomington: Indiana University.
MOSTEC [Ministry of Science, Technology, Education and Culture]. 2000. *Mongolia education sector strategy 2000–2005*. Ulaanbaatar: MOSTEC.
———. 2001. *School and dormitory buildings in Mongolia*. Compilation of MOSTEC for the donor meeting in Paris, May. Ulaanbaatar: MOSTEC.
———. 2003. *Dormitory enrollment figures*. Information sheet. Ulaanbaatar: MOSTEC.
MOSTEC [Ministry of Science, Technology, Education and Culture], UNDP, UNESCO, UNICEF, UNFPA, and World Bank. 2000. *Mongolia national report on Education for All assessment—2000*. Ulaanbaatar: MOSTEC.
Mostertz, W. 1982. Die Bedeutung des Beitritts der MVR zum Rat für Gegenseitige Wirtschaftshilfe für die Entwicklung der Volkswirtschaft. [The importance of the MPR joining to the Council for Mutual Economic Assistance for the development of the economy]. In *Die Mongolische Volksrepublik. Historischer Wandel in Zentralasien*, edited by author collective, 142–169. Berlin: Dietz-Verlag.
Mundy, K. 1999. Educational multilateralism in a changing world order: Unesco and the limits of the possible. *International Journal of Educational Development* 19: 27–52.
Mundy, K., and L. Murphy. 2001. Transnational advocacy, global civil society? Emerging evidence from the field of education. *Comparative Education Review* 45 (1): 85–126.
Nansal, R. 1971. BNMAU-yn üildverlekh khüchin üildverleliin khariltsaany tüükhen khögjliin zarim asuudal. [A few historical development problems regarding the relation between production forces and production]. In *BNMAU-yn kapitalist bus khöggliin tüükhen turshlagyn zarim asuudal*. Ulaanbaatar. [Editor and publisher not noted].
Narangoa, L. 1998. *Japanische Religionspolitik in der Mongolei 1932–1945. Reformbestrebungen und Dialog zwischen japanischem und mongolischem Buddhismus*. [Japan's religion policy in Mongolia 1932–1945. Reform initiatives and dialogue between Japanese and Mongolian Buddhism]. Studies in Oriental Religions, volume 43. Wiesbaden: Harassowitz.
Narmandakh, Ya. 1999. *Bolovsrolyn salbaryn khöggliin tösöl; deed bolovsrolyn chanaryg saijruulakh; khamtyn ajillagaag bekhjüülekh: Bagsh naryn mergejil deeshlüülekh*. [Education sector cooperation project, enhancing the quality of higher education and strengthen the cooperation: Professional development of teachers]. Ulaanbaatar: Asian Development Bank.
National Board for Children, Save the Children UK, and UNICEF. 2003. *The living conditions of the children in peri-urban areas of Ulaanbaatar*. Ulaanbaatar: Sogoo nuur publisher.
National Statistical Office of Mongolia. 2001. *2000 Population and housing census: The main results*. Ulaanbaatar: National Statistical Office of Mongolia.
———. 2002. *Gender in Mongolia: Analysis based on the 2000 census*. Ulaanbaatar: National Statistical Office of Mongolia.
National Statistical Office of Mongolia and World Bank. 2001. *Mongolia participatory living standards assessment*. Ulaanbaatar: National Statistical Office.

Natsagdorj, Sh. 1967. The economic basis of feudalism in Mongolia. *Modern Asian Studies* 1 (3): 265–281.

———. 1978. Das internationalistische Bündnis der werktätigen Araten mit der Arbeiterklasse des siegreichen Sozialismus—ein entscheidender Faktor der nichtkapitalistischen Entwicklung der MVR. [The internationalist alliance between the working arats and the working class of the triumphant socialism—a decisive factor of the non-capitalist development of the MPR]. In *Der revolutionäre Weg der Mongolischen Volksrepublik zum Sozialismus. Probleme der Umgehung des kapitalistischen Entwicklungsstadiums*, edited by H.-P. Vietze, 33–43. Studien über Asien, Afrika und Lateinamerika, Band 31. Berlin: Akademie-Verlag.

Nguyen, L. 1978. Grussansprache des Leiters der DRV-Delegation. [Key address of the chair of the delegation from the Democratic Republic of Vietnam]. In *Der revolutionäre Weg der Mongolischen Volksrepublik zum Sozialismus. Probleme der Umgehung des kapitalistischen Entwicklungsstadiums*, edited by H.-P. Vietze, 13–14. Studien über Asien, Afrika und Lateinamerika, Band 31. Berlin: Akademie-Verlag.

Njanday. 1976. Die produktive Arbeit der Schüler in der Mongolischen Volksrepublik. [The productive work of students in the Mongolian People's Republic]. *Polytechnische Bildung und Erziehung* 18 (5): 159–161.

Nolan, P. 1995. *China's rise, Russia's fall*. New York: St. Martin's Press.

Norovsambuu, S. 1971. Kapitalist bus khögjliin zamyn tukhai oilgoltyg todorkhoilokh asuudald. [Clarifying questions regarding the meaning of non-capitalist development]. In *BNMAU-yn Kapitalist bus khögjliin tüükhen turshlagyn zarim asuudal*, edited by Sh. U. A.-iin Tüükhiin Khüreelen, 7–40. Ulaanbaatar. [Publisher not noted].

Nóvoa, A., and M. Lawn. 2002. Introduction. In *Fabricating Europe: The formation of an education space*, edited by A. Nóvoa, and M. Lawn, 1–13 Dordrecht: Kluwer.

O'Day, J. A. 2002. Complexity, accountability, and school improvement. *Harvard Educational Review* 72 (3): 293–329.

Odgaard, O. 1996. Living standards and poverty. In *Mongolia in transition. Old patterns, new challenges*, edited by O. Bruun, and O. Odgaard, 103–134. Studies in Asian Topics, no. 22, Nordic Institute of Asian Studies, Richmond: Curzon Press.

Otgonjargal, O. 2004. *Higher education financial reform in Mongolia: Introduction of cost-sharing and its reasons, process and consequences*. Presentation at the Conference of the Mongolia Society, October. Bloomington: University of Indiana. Available at: <http://www.opensocietyforum.mn>. [Accessed on March 2, 2005].

Pacurari, O., B. Batkhuyag, and L. Mason. 2004. *Education sub-sector review: Teacher In-service training in Romania*. Course paper International Education Policy Studies, fall 2004. New York: Teachers College, Columbia University.

Parliament of Mongolia. 2002. Public Sector Management and Finance Act, signed June 27. Ulaanbbaatar: Parliament of Mongolia.

Pelzhee, M. 1981. An internatioal family. In *The 60th Anniversary of People's Mongolia*, 146–149. [Name of editor not noted]. Ulaanbaatar: Unen Editorial Board, and Moscow: Novosti Press Agency Publishing House.

Phillips, D. 2004. Toward a theory of policy attraction in education. In *The global politics of educational borrowing and lending*, edited by G. Steiner-Khamsi, 54–67. New York: Teachers College Press.

Picht, H. 1984. *Asien. Wege zu Marx und Lenin*. [Asia. Paths to Marx and Lenin]. Berlin: Dietz.

Pollack, E. 1993. Isaac Leon Kandel (1881–1965). *Prospects* 3 (4): 775–787.
Popkewitz, T. S. 1998. *Struggling for the soul: The politics of schooling and the construction of the teacher*. New York: Teachers College Press.
———. 2000. Globalization/regionalization, knowledge, and the educational practices: Some notes on comparative strategies for educational research. In *Educational knowledge*, edited by T. S. Popkewitz, 3–27. Albany: State University of New York Press.
Potkanski, T. 1993. Decollectivisation of the Mongolian pastoral economy (1991–1992): Some economic and social consequences. *Nomadic Peoples* 33: 123–135.
Pratt, N. 2003. On Martyn Hammersley's critique of Bassey's concept of the fuzzy generalisation. *Oxford Review of Education* 29 (1): 27–32.
Przeworski, A., and H. Teune. 1970. *The logic of comparative social inquiry*. New York: Wiley.
Ragin, C. C. 1997. Turning the tables: How case-oriented research challenges variable-oriented research. *Comparative Social Research* 16: 27–42.
Ramirez, F. O. 2003. The global model and national legacies. In *Local meanings, global schooling. Anthropology and world culture theory*, edited by K. Anderson-Levitt, 239–254. New York: Palgrave Macmillan.
Ramirez, F. O., and J. W. Meyer. 2002. National curricula: World models and national historical legacies. In *Internationalisierung—Internationalisation*, edited by M. Caruso, and H.-E. Tenorth, 91–107. Frankfurt/M: Lang.
Rasanayagam, J. 2003. *Market, state and community in Uzbekistan. Reworking the concept of the informal economy*. Halle, Germany: Max Planck Institute for Social Anthropology.
Rathmann, L., and H.-P. Vietze. 1978. Die Bedeutung des revolutionären Weges der MVR zum Sozialismus für die internationale Klassenauseinandersetzung. [The importance of the revolutionary path of the MPR towards socialism for the international class struggle]. In *Der revolutionäre Weg der Mongolischen Volksrepublik zum Sozialismus. Probleme der Umgehung des kapitalistischen Entwicklungsstadiums*, edited by H.-P. Vietze, 325–357. Studien über Asien, Afrika und Lateinamerika, Band 31. Berlin: Akademie-Verlag.
Rättig, U. 1974. Einige Bemerkungen zu den Schulen in der Zeit der Autonomie der Mongolei (1911–1920). [A few comments on schools during the autonomy period of Mongolia (1911–1920)]. In *Sprache, Geschichte und Kultur der altaischen Völker*, 489–495. Protokollband der XII. Tagung der Permanent International Altaistic Conference 1969 in Berlin. [Publisher not noted].
Ray, R. 1998. *Development economics*. Princeton: Princeton University Press.
Riegel, K.-G. 1993. Säuberungsriten in Virtuosengemeinschaften. Die Parteiversammlung deutschsprachiger Exilschriftsteller in Moskau (4.-8.9.1936). [Cleansing rituals in communities of virtuoso. The party congregation of exiled German writers in Moscow (September 4–8, 1936)]. *Kölner Zeitschrift für Sozialpsychologie* 33: 331–349.
Rinchen, B. 1957. Schamanistische Geister der Gebirge Dörben agula-yin ejed in Urgaer Pantomimen. [Shaman spirits of the Dörben agula-yin ejed mountain range in pantomines of Urga]. *Acta Ethnographica Hungarica* 16: 34–46.
———. 1964. *Mongol bichgiin khelnii züi. Tergüün devter.* [The science of written Mongolian language. Volume 1]. Ulaanbaatar: Shinjlekh Ukhaany Akademiin khevlekh üildver.

Rinchin, L. 1981. *Collectivisation—the key to a rise in the standard of living*. In *The 60th Anniversary of People's Mongolia*, 73–78. [Name of editor not noted]. Ulaanbaatar: Unen Editorial Board, and Moscow: Novosti Press Agency Publishing House.

Rogers, E. M. 1995. *Diffusion of innovations*. New York: Free Press (4th edition).

Rose, N. 1998. *Inventing our selves. Psychology, power, and personhood*. Cambridge: Cambridge University Press.

Rosen, L. 2003. The politics of identity and the marketization of U.S. schools: How local meanings mediate global struggles. In *Local meanings, global schooling. Anthropology and world culture theory*, edited by K. Anderson-Levitt, 161–182. New York: Palgrave Macmillan.

Rossabi, M. 1988. *Khubilai Khan. His life and times*. Berkeley and Los Angeles, CA: University of California Press.

———. 2005. *Modern Mongolia. From khans to commissars to capitalists*. Berkeley and Los Angeles: University of California Press.

Rottier, P. 2003. The Kazakness of sedentarization: Promoting progress as tradition in response to the land problem. *Central Asian Survey* 22 (1): 67–81.

Sachs, J. D. 2005. *The end of poverty. Economic possibilities for our time*. New York: Penguin.

Sachsenmeier, P. 1978. *Reformkonzeptionen der Lehrerbildung in der Sowjetunion*. [Teacher education reform concepts in the Soviet Union]. Weinheim and Basel: Beltz.

Samoff, J. 1999. Education sector analysis in Africa: Limited national control and even less national ownership. *International Journal of Educational Development* 19 (4/5): 249–272.

Sander, T., and Ya. Narmandakh. 1998. *In-service teacher education: Report to the Asian Development Bank*. Ulaanbaatar: MOSTEC.

Sandshaasüren, R., and I. Shernossek. 1981. *Das Bildungswesen der Mongolischen Volksrepublik. Beiträge zur Pädagogik, Band 22*. [The educational system in the Mongolian People's Republic. Contributions of education, volume 22]. Berlin: Volk und Wissen Volkseigener Verlag.

Sanjdorj, M. 1971. Ardyn tör bol kapitalist bus khögjliin gol zevseg mön. [The people's state is the main weapon for non-capitalist development]. In *BNMAU-yn kapitalist bus khögjliin tüükhen turshlagyn zarim asuudal*, 70–89. Ulaanbaatar. [Publisher not noted].

———. 1978. Zur Erforschung aktueller Probleme der nichtkapitalistischen Entwicklung der MRV. [An examination of current problems of the non-capitalist development of the MPR]. In *Der revolutionäre Weg der Mongolischen Volksrepublik zum Sozialismus. Probleme der Umgehung des kapitalistischen Entwicklungsstadiums*, edited by H.-P. Vietze, 471–483. Studien über Asien, Afrika und Lateinamerika, Band 31. Berlin: Akademie-Verlag.

Sanzhasuren, R. 1981. The torch of knowledge. In *The 60th Anniversary of People's Mongolia*, 89–91. [Name of editor not noted]. Ulaanbaatar: Unen Editorial Board, and Moscow: Novosti Press Agency Publishing House.

Schick, A. 1998. Why most developing countries should not try New Zealand's reforms. *World Bank Research Observer* 13: 123–132.

Schinkarjow, L. 1981. *Abschied von der Jurte. Erlebnisse in der Mongolischen Volksrepublik*. [Farewell from the yurt. Personal experiences in the Mongolian People's Republic]. Leipzig: VEB F. A. Brockhaus Verlag.

Schluss, J. H., and E. Sattler. 2001. Transformation—einige Gedanken zur Adaption eines nicht einheimischen Begriffs. [Transformation—a few thoughts on the adaptation of a non-indigenous term]. *Vierteljahresschrift für wissenschaftliche Pädagogik* 2: 173–188.

Schmidt, S. 1995. *Mongolia in transition: The impact of privatization on rural life*. Saarbrücken, Germany: Bielefelder Studien für Entwicklungssoziologie.

Schöne, U. 1973. *Die Entwicklung des Volksbildungswesens in der Mongolischen Volksrepublik 1921–1971. Ein Beitrag zur Analyse der Grundprobleme der Kultur- und Bildungsrevolution in Asien*. [The development of the educational system in the Mongolian People's Republic 1921–1971: A contribution to the analysis of the fundamental problems related to the cultural and educational revolution in Asia]. Dissertation. Berlin: Humboldt University.

———. 1982. Die Entwicklung des Volksbildungswesens in der Mongolischen Volksrepublik. [The development of the educational system in the Mongolian People's Republic]. In *Die Mongolische Volksrepublik. Historischer Wandel in Zentralasien*, edited by author collective, 170–184. Berlin: Dietz-Verlag.

———. 1988. *Einige Bemerkungen zu den mongolischen Schulen während der Mandschurenzeit*. [A few comments on Mongolian schools during the Manchu period]. Paper presented at the Permanent International Altaistic Conference (PIAC).

———. 1997. *20-iod onuudad German ulsad suraltsaj baisan mongol suragchdad bolon Laiptsigt khevlüülsen Mongolyn gazryn zurgyn tüükhend kholbogdoj zarim temdeglel*. [A few comments on the history of Mongolian students who studied in Germany in the 1920s, and on maps of Mongolia printed in Leipzig]. Paper presented at the Seventh International Congress of Mongolists in Ulaanbaatar.

Schriewer, J. 1990. The method of comparison and the need for externalization: Methodological criteria and sociological concepts. In *Theories and methods in comparative education*, edited by J. Schriewer, in cooperation with B. Holmes, 3–52. Bern: Lang.

———. 2000. World-system and interrelationship networks: The internationalization of education and the role of comparative inquiry. In *Educational knowledge: Changing relationships between the state, civil society, and the educational community*, edited by T. S. Popkewitz, 305–343. Albany: State University of New York Press.

Schriewer, J., and C. Martinez. 2004. Constructions of internationality in education. In *The global politics of educational borrowing and lending*, edited by G. Steiner-Khamsi, 29–53. New York: Teachers College Press.

Schriewer, J., J. Henze, J. Wichmann, P. Knost, S. Barucha, S., and J. Taubert. 1998. Konstruktion von Internationalität: Referenzhorizonte pädagogischen Wissens im Wandel gesellschaftlicher Systeme (Spanien, Sowjetunion/Russland, China). [The construction of internationality: Reference horizons of educational knowledge and social change]. In *Gesellschaften im Vergleich. Forschungen aus Sozial-und Geschichtswissenschaften*, edited by H. Kaelble and J. Schriewer, 151–258. Frankfurt/M: Lang.

Seddon, T. 2005. Traveling policy in post-socialist education. *European Educational Research Journal* 4 (1): 1–4.

Shagdar, S. 2000. *Mongol ulsyn bolovsrolyn tüükhiin tovchoon*. [A historical overview of the development of education in Mongolia]. Ulaanbaatar. [Publisher not noted].

Shagdarsüren, L. 1976. *BNMAU-yn yerönkhii bolovsrolyn khödölmör politekhnik surguuliin üüsel, khöggjilt*. [Origin and development of the general polytechnic school in the MPR]. Ulaanbaatar: Ardyn Bolovsrolyn Yaamny khevlel.

Sharkhüü, Ts. 1965. *Manjiin daranguillyn üyein Mongolyn surguul'* (1776–1891). [Mongolian schools during Manchu oppression (1776–1891)]. Ulaanbaatar: Ulsyn khevleliin khereg erkhlekh khoroo.
Sh. U. A. T. Kh. [Shinjlekh Ukhaany Akadem-iin Tüükhiin Khüreelen]. 1971. *BNMAU-yn Kapitalist bus khögjliin tüükhen turshlagyn zarim asuudal.* [A few issues regarding the history of non-capitalist development of the MPR]. Ulaanbaatar: Sh. U. A. T. Kh.
Shinjlekh Ukhaan, Bolovsrolyn, Khün Amyn Bodlogo, Khödölmöriin, Sangiin Said. [Ministers of Education and Science, Demographics, Labor, and Finance]. 1995. *Tsetserleg, surguuliin udirdakh ajiltan, bagsh, surgan khümüüjüülegchded mergej liin zereg, ur chadvariin nemegdel olgon juram.* [Regulation on giving professional titles, qualification- and skills-bases bonuses to administrative employees and teachers in kindergarten and in secondary schools], signed December 26. Ulaanbaatar: Ministry of Education and Science.
Shirendev, B. 1967. *Kapitalismyg algasaad.* [Bypassing capitalism]. Ulaanbaatar. [Publisher not noted].
Shirendyb, B. 1971. *Die Mongolische Volksrepublik. Von der Feudalordnung in den Sozialismus.* [The Mongolian People's Republic. From feudal order to socialism]. Berlin: Staatsverlag der DDR.
———. 1978. Einige Probleme aus der Geschichte der nichtkapitalistischen Entwicklung der MVR zum Sozialismus. [A few problems from MPR's history of non-capitalist development towards socialism]. In *Der revolutionäre Weg der Mongolischen Volksrepublik zum Sozialismus. Probleme der Umgehung des kapitalistischen Entwicklungsstadiums*, edited by H.-P. Vietze, 15–32. Studien über Asien, Afrika und Lateinamerika, Band 31. Berlin: Akademie-Verlag.
———. 1981. A historic choice. In *The 60th Anniversary of People's Mongolia*, 19–25. [Name of editor not noted]. Ulaanbaatar: Unen Editorial Board, and Moscow: Novosti Press Agency Publishing House.
Silova, I. 2004. Adopting the language of the new allies. In *The global politics of educational borrowing and lending*, edited by G. Steiner-Khamsi, 75–87. New York: Teachers College Press.
———. 2005. *From sites of occupation to symbols of multiculturalism: Transfer of global discourse and the metamorphosis of Russian schools in post-Soviet Latvia*, Greenwich, CT: Information Age Publishing.
Silova, I., and G. Steiner-Khamsi. 2005. *Dealing with the postsocialist reform package: From Baku to Ulaanbaatar.* Roundtable at the Annual Conference of the International Comparative Education Society, March 22–26. Stanford: Stanford University School of Education.
Skutnabb-Kangas, T. 2000. *Linguistic genocide in education, or worldwide diversity and human rights?* Mahwah, NJ: L. Erlbaum Associates.
Smyth, J., and A. Dow. 1998. What's wrong with outcomes? Spotter planes, action plans, and steerage of the educational workplace. *British Journal of Sociology of Education* 19 (3): 291–303.
Sneath, D. 2002a. Mongolia in the "Age of the Market": Pastoral land-use and the development discourse. In *Markets and moralities. Ethnographies of postsocialism*, edited by R. Mandel and C. Humphrey, 191–210. Oxford and New York: Berg.
———. 2002b. Reciprocity and notions of corruption in contemporary Mongolia. *Mongolian Studies* 25: 85–99.
———. 2003. Lost in the post: Technologies of imagination, and the Soviet legacy in post-socialist Mongolia. *Inner Asia* 5: 39–52.

Sodnomgombo, D. 1978. Die grundlegenden Veränderungen in der Klassenstruktur der MVR als Ergebnis der Überwindung der sozialen Widersprüche in der Übergangsperiode vom Feudalismus zum Sozialismus. [The fundamental change of class structure in the MPR as a result of having resolved the social contradiction during the transition period from feudalism to socialism]. In *Der revolutionäre Weg der Mongolischen Volksrepublik zum Sozialismus. Probleme der Umgehung des kapitalistischen Entwicklungsstadiums*, edited by H.-P. Vietze, 95–112. Studien über Asien, Afrika und Lateinamerika, Band 31. Berlin: Akademie-Verlag.

Songino, Ts. 1991. *Mongolyn belcheeriin mal mallagaany ulamjlal* [Traditions in Mongolian animal husbandry]. Ulaanbaatar: "Soyombo" khevleliin gazar.

Soros Foundation Kyrgyzstan. 2004. *Proposal for implementation of a voucher mechanism in education financing*. Bishkek, Kyrgyz Republic: Soros Foundation Kyrgyzstan.

Spaulding, S., T. Boldsukh, D. Munkjargal, and O. Otgonjargal. (1999). *Improvement of educational management and supervision for graduate preparation programs in educational administration: Report to the ADB*. Ulaanbaatar: MOSTEC.

Spreen, C. A. 2004. Appropriating borrowed policies: Outcomes-based education in South Africa. In *The global politics of educational borrowing and lending*, edited by G. Steiner-Khamsi, 101–113. New York: Teachers College Press.

Stambach, A. 2003. World-cultural and anthropological interpretations of "choice programming" in Tanzania. In *Local meanings, global schooling. Anthropology and world culture theory*, edited by K. Anderson-Levitt, 141–160. New York: Palgrave Macmillan.

Steiner-Khamsi, G. 2001. *Year 3 School 2001 Evaluation Development Program. On-site evaluation in the provinces Khovd and Bayan-Ölgii*. Report of August 13. Ulaanbaatar: Mongolian Foundation for Open Society.

———. 2002. Reterritorializing educational import: Explorations into the politics of educational borrowing. In *Fabricating Europe. The formation of an education space*, edited by A. Nóvoa and M. Lawn, 69–86. Dordrecht: Kluwer.

———. 2003. Vergleich und Subtraktion: Das Residuum zwischen Globalem und Lokalem. [Comparison and subtraction: The residual between the global and the local]. In *Vergleich und Transfer. Komparatistik in den Sozial-, Geschichts- und Kulturwissenschaften*, edited by H. Kaelble and J. Schriewer, 369–397. Frankfurt/M: Campus.

Steiner-Khamsi, G., ed. 2004a. *The global politics of educational borrowing and lending*. New York: Teachers College Press.

Steiner-Khamsi, G. 2004b. Blazing a trail for policy theory and practice. In *The global politics of educational borrowing and lending*, edited by G. Steiner-Khamsi, 201–220. New York: Teachers College Press.

Steiner-Khamsi, G., ed. 2005a. *Bodlogod nölöölökh bolovsrolyn sudalgaa*. [Policy-relevant educational research]. Ulaanbaatar: Mongolian Education Alliance Publisher.

Steiner-Khamsi, G. 2005b. Vouchers for teacher education (non) reform in Mongolia: Transitional, postsocialist, or antisocialist explanations? *Comparative Education Review* 49 (2): 148–172.

Steiner-Khamsi, G., and A. Gerelmaa. 2005. *Public Expenditure Tracking Survey interim report*. Washington, DC: World Bank, and Ulaanbaatar: Open Society Forum.

Steiner-Khamsi, G., and O. Kuliyash. 2004. Angiin darga. [The class monitor]. *Bolovsrol Sudlal* 17 (5): 66–81.

Steiner-Khamsi, G., and A. T. Nguyen. 2001. *Seasonal and permanent migration in Mongolia: A preliminary assessment of access and quality of education* (unpublished sector note). Washington, DC: World Bank.
Steiner-Khamsi, G., and H. O. Quist. 2000. The politics of educational borrowing: Reopening the case of Achimota in British Ghana. *Comparative Education Review* 44 (3): 272–299.
Steiner-Khamsi, G., and I. Stolpe. 2004. De- and recentralization reform in Mongolia: Tracing the swing of the pendulum. *Comparative Education* 40 (1): 29–53.
———. 2005. Non-traveling "best practices" for a traveling population: The case of nomadic education in Mongolia. *European Educational Research Journal* 4 (1): 22–35.
Steiner-Khamsi, G., O. Myagmar, and B. Sum"yaasüren. 2004. Surgan khümüüjüülekh shog yaria ba khicheel zokhion baiguulalt. [Pedagogical jokes and classroom organization and management]. *Lavai* 1 (1): 8–20.
Steiner-Khamsi, G., I. Silova, and E. Johnson. 2006. Neoliberalism liberally applied: Educational policy borrowing in Central Asia. In *2006 World Yearbook on Education*, edited by, D. Coulby, J. Ozga, T. Seddon, and T. S. Popkewitz, 217–245. London and New York: Routledge.
Steiner-Khamsi, G., S. I. Stolpe, and Tümendelger, 2003. Bolovsrolyn tölöökh nüüdel [School-related migration]. *Shine Tol'* 45 (4): 82–112.
Steiner-Khamsi, G., I. Stolpe, and A. Gerelmaa. 2004a. *Rural School Development Project in Mongolia. Evaluation report*. Ulaanbaatar and Copenhagen: Danish Mongolia Society and Mongolian Association for Primary and Secondary School Development in collaboration with Copenhagen International Centre for Educational Development (CICED) and Mongolian State University of Education.
———. 2004b. *Rural School Development Project in Mongolia. Transcripts of interviews*. New York, Berlin, Ulaanbaatar: Unpublished.
Steiner-Khamsi, G., D. Tümendemberel, and E. Steiner. 2005. Bagsh mergejiltei etseg ekhchüüd. [Teachers as parents]. *Bolovsrol Sudlal* 18 (1): 40–53, 18 (2): 62–70.
Steiner-Khamsi, G., N. Enkhtuya, T. T. Prime, and S. R. Lucas. 2000. *School 2001 Education Development Program. Evaluation report year 2*. Ulaanbaatar: Mongolian Foundation for Open Society.
Stiglitz, J. E. 2003. *Globalization and its discontents*. New York: Norton.
Stolpe, I. 2001. *Zur Transformation des Grund- und Sekundarschulwesens in der Mongolei von 1990–1999*. [The transformation of basic and secondary education in Mongolia, 1990–1999]. Masters thesis. Berlin: Humboldt University.
———. 2003. Erschaffung eines Drittweltlandes: Nomadenbildung in der Mongolei. [The creation of a Third World country: Education of nomads in Mongolia]. *Tertium Comparationis* 9 (2): 162–177.
Sükhbaataryn neremjit khevleliin kombinat [Sükhbaatar publishing association]. 1964. *BNMAU-yn kapitalist bish zamyn zarim züi togtool ontsloguud ba tus ulsyn khögjliin khetiin tölövöös*. [A few rules and special features of the non-capitalist development of the MPR, and future perspectives of this state]. Ulaanbaatar: Sükhbaataryn neremjit khevleliin kombinat.
Szynkiewicz, S. 1993. Mongolia's nomads build up a new society again: Social structure and obligations on the eve of the private economy. *Nomadic Peoples* 33: 163–172.
Taube, E., and M. Taube. 1983. *Schamanen und Rhapsoden. Die geistige Kultur der alten Mongolei*. [Shamans and rhapsodists. The spiritual culture of ancient Mongolia]. Leipzig: Koehler & Amelang.

Tichomirov, V. D. 1978. Einige Probleme der ökonomischen Zusammenarbeit der MVR mit den sozialistischen Staaten. [A few problems of the economic collaboration of the MPR with the socialist states]. In *Der revolutionäre Weg der Mongolischen Volksrepublik zum Sozialismus. Probleme der Umgehung des kapitalistischen Entwicklungsstadiums*, edited by H.-P. Vietze, 407–417. Studien über Asien, Afrika und Lateinamerika, Band 31. Berlin: Akademie-Verlag.

Tilly, C. 1997. Means and ends of comparison in macrosociology. *Comparative Social Research* 16: 43–53.

———. 2004. Past, present, and future globalizations. In *The global politics of educational borrowing and lending*, edited by G. Steiner-Khamsi, 13–28. New York: Teachers College Press.

Todd, A., and M. Mason. 2005. Enhancing learning in South African schools: Strategies beyond outcomes-based education. *International Journal of Educational Development* 25: 221–235.

Tömörjav, M. 1999. Nutgiin mongol mal. Töv Aziin baigal', tsag uur, ekologiin nökhtsöld dasan zokhison baidal. [How the Mongolian livestock is adapted to the nature, climate and ecology of Central Asia]. *Mongolica* 9: 337–349.

Tömörjav, M., and N. Erdenetsogt. 1999. *Mongolyn Nüüdelchin*. [The Mongolian nomad]. Ulaanbaatar: Mongolyn shinjlekh ukhaany akademi, nüüdliin soyol irgenshliig sudlakh olon ulsyn khüreelen.

Tsedenbal, Yu. 1954. *MAKhN-yn TKh-noos namyn 12-r ikh khurald tav'san iltgel ba tüünd gargasan togtool*. [Compilation of presentations and decisions from the 12th annual conference of the central committee of the MPRP]. Ulaanbaatar. [Publisher not noted].

Tsedenbal-Filatova, A. I. 1981. Everything for the happiness of children. In *The 60th Anniversary of People's Mongolia*, 104–108. [Name of editor not noted]. Ulaanbaatar: Unen Editorial Board, and Moscow: Novosti Press Agency Publishing House.

Tserensodnom, D. 2001. *Sar shiniin beleg*. [A gift of the new moon]. Ulaanbaatar: Öngot khevlel.

Tsevelmaa, Sh. 1965. Suragchdyg biye daalgan bodlogo boduulj surga"ya. [Let us promote the self-initiative of students]. *Surgan Khümüüjüülegch* (1): 31–36.

Tüdev, B. 1971. Ündesnii aj üildver, ajilchin angi üüsch khögjsön ni BNMAU-yn kapitalist bus khögjliin gol chukhal ür dün mön. [The emergence of a national industry and working class is one of the most important results of the non-capitalist development of the MPR]. In *BNMAU-yn kapitalist bus khögjliin tüükhen turshlagyn zarim asuudal*, 181–199. Ulaanbaatar. [Publisher not noted].

Tuyaatsetseg, B. 2001. Am'dralyg khöglökh ukhaan. [Life science]. *Khongorzul* (007/304): 3.

Tyack, D., and L. Cuban. 1995. *Tinkering toward utopia: A century of public school reform*. Cambridge, MA: Harvard University Press.

Uhlig, G. 1989. Zu den Anfängen der Elementarbildung in der Mongolei (1921–1940). [The origins of elementary education in Mongolia (1921–1940)]. *Vergleichende Pädagogik* 25: 403–409.

Ulziisaikhan, G. 2004. *On effects of loan and aid in rural development*. Available at: <http://www.opensocietyforum.mn>. [Accessed March 14, 2005].

UNDP [United Nations Development Programme]. 2000. *Survey report on NGO implemented assistance in social sector of Mongolia*. Conducted jointly by Consulting and Business Centre, Academy of Management, Gender Centre for Sustainable Development, Consulting Unit zbn. Ulaanbaatar: UNDP.

———. 2001. *Second country cooperation framework for Mongolia (2002–2006)*. Executive Board of the United National Development Programme and the United Nations Population Fund, session of September 10–14, 2001. United Nations, UNDP: New York.

———. 2005. *The CEE/CIS/Baltics region*. Available at: <http://www.undp.org/regions/europe/>. [Accessed February 28, 2005].

UNICEF [United Nations Children's Fund]. 1999. *After the fall. The human impact of ten years of transition*. Florence, Italy: UNICEF, Innocenti Research Centre.

———. 2000. *Children and women in Mongolia. Situation analysis report 2000*. Ulaanbaatar: UNICEF.

———. 2005. *Girl's Education. Regional perspectives: East Asia and the Pacific*. New York: UNICEF. Available at: <http://www.unicef.org/girlseducation/index_regionalperspectives.html>. [Accessed February 25, 2005].

UNIFEM [United Nations Development Fund for Women] and UNDP [United Nations Development Programme]. 2002. *A gender lens on the rural map of Mongolia: Data for policy*. Ulaanbaatar: UN Building.

Unkrig, W. A. 1929. *Das Programm des Gelehrten Comités der Mongolischen Volksrepublik*. [The program of the scholar's committee of the Mongolian People's Republic]. Berlin: Mitteilungen des Seminars für Orientalische Sprachen zu Berlin, Jahrgang XXXII, Abteilung I, Ostasiatische Studien.

Urtnasan, N. 1991. Bolovsrolyn asuudal. [Educational issues]. *Ardyn Erkh* November 11: 2.

Vavrus F. 2003. Desire and decline: Schooling amid crisis in Tanzania. New York: Lang.

———. 2004. The referential web: Externalization beyond education in Tanzania. In *The global politics of educational borrowing and lending*, edited by G. Steiner-Khamsi, 141–153. New York: Teachers College Press.

Veit, V. 1985. Das Pferd—Alter Ego des Mongolen? Überlegungen zu einem zentralen Thema der mongolischen Geschichte und Kultur. [The horse—alter ego of Mongol? A few thoughts on a central theme in Mongolian history and culture]. *Asiatische Forschungen* (91): 58–88.

Verdery, K. 1996. *What was socialism and what comes next?* Princeton: Princeton University Press.

———. 2002. Whither postsocialism? In *Postsocialism. Ideals, ideologies and practices in Eurasia*, edited by C. M. Hann, 15–21. London and New York: Routledge.

———. 2003. *The vanishing hectare: Property and value in postsocialist Transylvania*. Ithaca, NY: Cornell University Press.

Vietze, H.-P., ed. 1978. *Der revolutionäre Weg der Mongolischen Volksrepublik zum Sozialismus. Probleme der Umgehung des kapitalistischen Entwicklungsstadiums*. [The revolutionary path to socialism of the Mongolian People's Republic. Problems regarding bypassing the capitalist development stage]. Studien über Asien, Afrika und Lateinamerika, Band 31. Berlin: Akademie-Verlag.

Vladimirtsov, B. Ya. 1934. *Obshchestvenny stroi mongolov. Mongolskii kochevoj feodalizm*. [(Russian) The social order of Mongolians. The Mongolian nomadic feudalism]. Leningrad: Izdatelstvo Akademii Nauk SSSR.

Voslensky, M. S. 1984. *Nomenklatura. Die herrschende Klasse der Sowjetunion*. [Nomenclatura. The ruling class of the Soviet Union]. München: Molden.

Wallerstein, I. 1974. *The Modern world-system I. Capitalist agriculture and the origins of the European world-economy in the sixteenth century*. New York: Academic Press.

———. 2004. *World-systems analysis. An introduction*. Durham and London: Duke University Press.

Watts, D. J. 2003. *Six degrees. The science of a connected age*. New York: Norton.

Weidman, J. 2001. Developing the Mongolia education sector strategy 2000–2005: Reflections of a consultant at the Asian Development Bank. *Current Issues in Comparative Education* 3 (2), online journal. Available at: <http://www.tc.edu/cice>. [Accessed November 24, 2005].

Weidman, J. C., J. L. Yeager, R. Bat-Erdene, J. Sukhbaatar, Ts. Jargalmaa, and S. Davaa. 1998. Mongolian higher education in transition: Planning and responding under conditions of rapid change. *Tertium Comparationis* 4 (2): 75–90.

Wolff, S. 1971. Mongolian educational venture in Western Europe (1926–1929). *Zentralasiatische Studien* 5: 247–320.

World Bank. 1992. *Mongolia. Toward a market economy*. Washington, DC: World Bank.

———. 2001. *Mongolia country assistance evaluation*. Washington, DC: World Bank, Operations Evaluation Department.

———. 2002. *Public expenditure and financial management review. Bridging the public expenditure management gap*. Washington, DC: World Bank.

———. 2004a. *Memorandum of the president of the International Development Association to the executive directors on a country assistant strategy of the World Bank Group for Mongolia*. April 5, 2004. Washington, DC: World Bank, Southeast Asia, and Mongolia Country Unit, East Asia and Pacific Region.

———. 2004b. *Debt initiative for heavily indebted poor countries. Annual meetings 2004*. Washington, DC: World Bank.

———. 2005. *Regional map of the Asia/Pacific region*. Available at: <http://www.worldbank.org>. [Accessed February 10, 2005].

Zlatkin, I. Ya., and M. I. Gol'man. 1982. Das Nomadenvolk der Mongolen auf dem historischen Weg vom Feudalismus zur Volksrevolution 1921. [The nomad people of the Mongolians on their historical path from feudalism to people's revolution 1921]. In *Die Mongolische Volksrepublik. Historischer Wandel in Zentralasien*, edited by author collective, 7–29. Berlin: Dietz-Verlag.

Zuunmod. 2004. *Zuunmod sumyn IV zakhirgaanaas bagsh . . . —tai baiguulsan "Ür düngiin gereee"-g dügnekh khüsnegt* [Schematic overview of the "outcomes-contract," agreed upon between the rural-district administration No. 4 from Zuunmod and teacher . . .]. Zuunmod: Rural district-school, Töv province.

1924 ony khevlel [Publisher of the year 1924]. *BNMAU-yn ankhdugaar ündsen khuul'*. [First constitutional law of the MPR]. [Place of publication not noted].

Index

2002 Education Law, 100–101
2004 Human development report, 93

Abu-Lughod, Janet, 23
Academy of Sciences, 40, 48, 95
agrarian-industrial state, 69, 192
aid-dependent, Mongolia as, 74
Akiba, M., 215*n*
Aklog, Fenot, 212*n*
Alexijewitsch, Swetlana, 54, 61
Amar, N., 92–93
Amgalan, Dagdangiin, 51–52, 55, 65–66, 207*n*
Anderson-Levitt, K., 6, 7, 200
Appadu, K., 86
Appadurai, Arjun, 65
Asian Development Bank (ADB), 74–77, 86–89, 104, 108, 136, 144, 161, 173, 176, 187, 214*n*, 215*n*
Atwood, Christopher, 11–12

Baabar, Bat-Erdeniin, 26, 30, 35, 37, 206*n*
Baasanjav, Z., 32, 206*n*
Baker, David, 124–125, 197
Baljirgarmaa, B., 36, 206*n*
Barkmann, Udo, 30, 33
Bawden, Charles, 27, 31, 61
Bayasgalan, S., 39
benchmarks, 80, 113, 131, 133, 136, 138, 212*n*
bilateral donors, 74, 77–78, 81, 194
boarding schools, 18, 42, 45, 70, 80, 92, 165–172, 175–177, 183, 199, 210*n*, 215*n*
Bogd Gegeen, 34, 40
Boldbaatar, J., 28
Boli, John, 125
Boone, Peter, 87

Brown versus Board of Education, 195
Buddhism and Buddhists, 11, 12, 25–29, 32, 34, 41, 165–166, 205*n*, 206*n*
  Lamaism, 25, 27, 28
Bulag, Uradyn, 13, 24, 30, 38–39, 54, 57, 60, 69
Buriats, 12, 35, 40, 206*n*
bypassing capitalism, 46, 51, 55–56, 58, 60, 63, 65, 67, 69, 72, 191
"Bypassing Capitalism" (painting), 51–53, 66

capitalism, 15, 51, 71. *See also* bypassing capitalism
Cascade Reform Strategy, 112
Chabbott, C., 8, 125
Child-Friendly School (CFS) program, 80
China, People's Republic of, 7, 11–12, 30, 32, 34, 36, 38, 69, 79, 81, 88–90, 127, 195, 205*n*, 208*n*, 215*n*
choice, 3, 6, 7, 9–10, 132, 135, 147, 149–152, 157, 198, 202
civil servants, 20, 31–34, 96, 98, 134, 136, 137, 160, 166, 198, 210*n*
class monitors, 19, 75, 110, 113–121, 141, 197–198, 210*n*, 211*n*
Cold War, 4, 6, 16–17, 60, 72, 77, 78, 195–196
collectivization of livestock farming, 13, 45, 62, 166, 170, 192, 215*n*
colonization, 6, 11, 12, 27, 29–34, 128, 188
Communist International (Comintern), 40, 41, 56, 206*n*, 207*n*
Complex Program, 68, 208*n*
complex schools, 103, 107–108, 118

convergence of student achievement outcomes, 2–3, 6–9, 185–186, 197, 202
Corvalan, Louis, 71
Council for Mutual Economic Assistance (CMEA), 2, 45, 67–71, 73, 87, 194, 207n, 215n
Cuba, 16, 72, 127, 194, 207n
Cuban, Larry, 155
cultural homogenization, 71
culture campaigns, 192, 193
Cummings, William K., 126–128, 197
Cyrillic alphabet, 42, 103, 149

Dalai Lama, 27, 34, 205n
Danish International Development Assistance program (DANIDA), 17–19, 77, 80, 94, 100, 112, 161, 182, 183, 215n
Davis, Angela, 71
decentralization of educational finance and governance, 99–102
decolonization, 37, 38, 67, 69
democracy, 62, 64, 65, 102, 147, 207n
Democratic Union, 73, 104, 150, 175, 208n
demographics, 1–2, 11–15
Desai, Padma, 86
"developing country" status, 67–68, 73
Dewey, John, 7, 128
Dimou, Augusta, 191
divergence of student achievement outcomes, 3–4
donor logic, 74, 77, 80, 82, 193
dormitory fees, 103
Dorzhsuren, Y., 42
Dow, A., 211n
dropout rates, 14, 169, 177, 179, 181–182, 193, 199
   statistics for, 177–182
dropouts, 2, 20, 82, 103, 117, 121, 122, 167, 169, 173–182, 193, 199, 214n

early adopters, 9–10, 134, 200
Education for All, 8, 40, 42, 44, 49, 80, 152, 171, 180, 182
education sector development projects, 76, 176, 208n

educational development compared with wrong amount of national income, 92–94
educational development compared with wrong countries, 88–92
educational import, four eras of, 26, 128, 188
   First Era, Enlightenment, 25–29, 49
   Fourth Era, Universal Access, 38–49
   Second Era, Colonization, 29–34
   Third Era, Nation Building, 34–38
educational spending, 88–89, 91–93, 209
   percentage from GDP, 88, 91–93
electrotechnics, 63
"embedded liberalism," 78
empirical studies, 19
Enkhtuya, Natsagdorj, 18, 213n
enlightenment, 25–29, 128, 188
enrollment ratio, 2, 73, 165
epidemic, reform, 9–10, 134, 182, 200
epochs of the postrevolutionary period, 61
Erdene-Batukhan, 36, 40, 41
Erdenejargal, Perenlei, 18

feudalism, 46, 51, 53, 55, 57–65, 67, 191, 192
"flag of convenience," 14, 144, 155, 157, 159, 161
Forman, Werner, 29
Foucault, M., 54
Frank, André Gunder, 11
Frederic, N., 86

gegeerel, 28–29
gender gap, reverse, 2, 121–123, 180, 198
"(generally) democratic," 61
generation gap, 2
geography of Mongolia, 11, 88–92
Gerelmaa, Amgaabazar, 18, 213n
Germany, 4, 16, 35, 41, 69, 74, 77, 125–126, 131, 189, 191, 206n, 207n, 211n
Giordano, Christian, 15, 190
girls, education of, 36, 42, 81, 157, 168, 180
global schooling, 6

INDEX 243

globalization, 1–8, 20, 23, 83, 123, 125, 131, 134, 135, 183, 185–186, 188, 191, 193, 196, 199–203
Goesling, B., 215n
Gorbachev, Mikhail, 208n
Great People's Khural, 43
Great Socialist October Revolution of 1917, 36
Gundsambuu, Khayankhyarvaagiin, 64–65

Hambly, Gavin, 30
Hardt, Michael, 15
"heavily indebted poor countries" (HIPD), 208n
Heyneman, S. P., 103
hierarchical settings in Mongolian classrooms, 115–116. *See also* class monitors
higher education, 14, 36, 48, 49, 116, 122, 189
  tuition-based, 94–99, 197, 212n
historiography, 58–59, 61, 69, 190, 192, 207n
Holmes, Brian, 77
Hopkin, A. G., 23
hPags-pa, 26–27
Human development report Mongolia, 66
Humphrey, Caroline, 13, 15–16, 63, 64, 65, 159, 170, 171
hybridization, 28, 32, 197, 201, 202
Hyer, Paul, 28, 32

ideoscapes, 65
illiteracy, 42, 59, 62, 70, 192, 193
industrial-agrarian state, 46, 63, 69, 192
Inner Mongolia, 11, 12, 26, 30–31, 37, 42, 60, 207n, 214n
International Association for the Evaluation of Educational Achievement (IEA), 19, 124, 199
international financial institutions, 74–77, 85. *See also* Asian Development Bank (ADB); International Monetary Fund (IMF); World Bank
international model of education, 2–3, 14
International Monetary Fund (IMF), 74, 78, 85–87, 209n, 214n

internationalism, 43, 59, 68–69, 71, 158
Islamic education, 127

Jagchid, Sechen, 28, 32
Japanese International Cooperation Agency (JICA), 78, 176
Jigmedsüren, S., 36, 206n
Jones, Phillip, 8, 80, 187

Kalinin, M. I., 46
Kausylgazy, N., 39, 49
Kazakhstan, 11, 13, 81, 89, 134, 189, 206n, 207n, 209n
Keuffer, Josef, 29
Khan, Altan, 27, 205n
Khan, Chinggis, 25, 51, 58, 205n
Khan, Khubilai, 25, 26, 27
Khazanov, Anatoly, 58, 171
Kim, Byung-Yeon, 94
Koeberle, S., 209n
Korea International Cooperation Agency (KOICA), 78
Kostova, Dobrinka, 15, 190
Kratli, Saverio, 45, 171–172
Krupskaya, N. K., 7, 46, 47
Kuliyash, Onorkhan, 19, 114
Kyrgyzstan, 11, 13, 89, 91, 134, 163, 207n

Lamaism (Tibetan Buddhism), 25, 27, 28. *See also* Buddhism and Buddhists
Lattimore, Owen, 30
Ledeneva, Alena, 93–94
legitimization strategies, 37–38
Lenin, Vladimir I., 7, 41, 55–56, 60–61, 120–121, 206n, 207n
LeTendre, Gerald, 124–125, 197, 215n
Lhagve, S., 149
lifelong learning, 147, 155–156
line monitors, 115, 119
literacy, 4, 26, 33, 42, 59, 73, 165, 174, 182, 193
Lkhagvajav, Ch., 39
Lkhagvasüren, G., 28
loose coupling between envisioned and enacted policies, 94, 196, 202
Lunacharskii, A. V., 46
Lynch, James, 157

Machel, Samara Moises, 80
Mandel, Ruth, 81–82, 159
Mao Tse-tung, 69, 215n
Martinez, C., 187
Marx, Karl, 53–56
Marxism, 39, 44, 46, 51, 53–65, 68, 156, 191–192, 206n, 207n
Marxist model of the historical stages of development, 65–66
meat requirement, 103, 105, 169, 175
Millennium Development Fund, 74, 77
Ministry for the People's Education, 40
modern schooling, 5, 125–129, 202
monastic schools, 25, 27–29, 32–33, 37, 40–41, 57, 165–166, 205n
Manchu Empire, 12, 30, 32
Manchuria, 12, 29–33, 129, 188, 206n
Meyer, John, 5, 6, 7, 125
migration, 2, 13, 19, 105, 149, 167, 170, 172, 174, 175, 205n, 214n
 destruction of, 41–42
"Mongol," 13
Mongolian Dropout Study, 173, 181
Mongolian Education Alliance (MEA), 82, 181, 205n, 213n, 215n
Mongolian Foundation for Open Society (MFOS), 17–19, 81–83, 109, 112, 158, 161–163, 194, 205n, 213n, 215n
Mongolian GDP
 accuracy of calculation, 92–94
 percentage of education expenditures from, 88, 91–93
 percentage of internationalist aid as, 73, 87, 209n
Mongolian People's Revolution in 1921, 38, 55
Mongolian People's Revolutionary Party (MPRP), 59, 73–74, 80, 107, 150, 154, 175–177, 208n, 213n
Mongolian State University, 42
Morgan, D., 26
Morozova, I., Y., 41
Moses, Larry, 27, 206n
Mozambique, 16, 80, 194
multi-institutionalist theory, need for, 123–129
Muslim education. *See* Islamic education
Myagmar, Ochirjav, 19, 119

Namnansüren, Sain Noyon Khan, 37
nation building, 34–38, 127, 128, 188
negdel (animal husbandry collectives), 45, 87, 105, 166, 170, 210n
Negri, Antonio, 15
neoinstitutionalism, 5–7, 124–126, 197, 202
neoliberalism, 2, 9, 10, 65, 103
New Accountability, 131, 133, 211n
New Contractualism, 131, 133, 136, 145, 198
New Public Management, 131, 134, 136–137
new teaching technologies. *See* student-centered learning
New Zealand, 90, 131–136, 142, 144, 200, 214n
Niislel Khüreenii sonin bichig, 36, 37
Njanday, 47
No Child Left Behind Act, 211n
nomadic education, 2, 45, 65, 165–183, 199
 best practices in, 171–172
 nomadic lifestyle, 13, 17, 149, 166
 seasonal migration and, 172–175
nomadic feudalism, 57–58
"nomadism," 24
non-enrollment, 103, 121–122, 175, 177
nongovernmental organizations (NGOs), 20, 74, 81–83, 89, 147, 150, 161–162, 198, 199, 215n

obligatory schooling, 41, 43, 44
O'Day, Jennifer, 211n
Open Society Forum, 83, 205n, 213n, 215n
Open Society Institute (Soros Foundation), 13, 89, 209
Organization for Economic Cooperation and Development (OECD), 4, 19, 88, 93, 199
outcomes-based education (OBE), 4, 6, 131–139, 141–145, 182, 187, 189, 196, 198, 200, 211n, 212n
outcomes-contracts, 131, 133, 136–137, 141, 143, 198, 212n
Outer Mongolia, 11, 12, 26, 30–31, 33–35, 38, 206n

pan-Mongolism, 12, 34, 37, 38, 40, 42
pedagogical humor, 110–113
"people's education," 40, 42–44, 49
Phillips, David, 131
pioneer camps, international,
　71, 208*n*
policy action, 9, 31, 44, 94, 154, 186
policy attraction, 3, 9, 131, 198
policy implementation, 9, 94, 152, 157,
　185, 186
policy talk, 9, 37, 44, 94, 147, 154,
　157, 186, 187, 202, 203
politics of transition, 65
polytechnic education, 26, 43, 45,
　46–47, 49, 77, 206*n*
postsocialism, 2, 5–6, 10, 13–18, 24,
　81–83, 86, 139, 156–157,
　169–171, 183, 188–193, 196, 198,
　200, 207*n*, 209*n*, 211*n*
"postsocialist reform package,"
　14, 189
postsocialist studies, 16, 54, 191
poverty, 45, 64–66, 77, 80, 86, 175,
　180, 182, 207*n*, 209*n*, 210*n*, 214*n*
Poverty Alleviation Program, 86, 175,
　182, 207*n*, 209*n*, 214*n*
Programme for International Student
　Assessment (PISA), 4
"public choice," 147, 149, 152
Public Expenditure Tracking Survey
　(PETS), 17, 144, 210*n*, 211*n*
Public Sector Management and Finance
　Act, 101, 136, 212*n*

Qing dynasty, 26, 29–32, 34, 206*n*

Ramirez, Francisco, 5–6, 7, 125, 202
rationalization of school staff, 103–108
Rättig, Uta, 36
recentralization law. *See* 2002
　Education Law
Red Hats, 27, 205*n*
reorganization of schools, 103–108
research, 17–21
"residuum," 24, 49
revisionism history, 58–59
Rinchen, Byamba, 27, 29, 32,
　33, 205*n*
Rossabi, Morris, 26, 208*n*
rote learning, 29, 49

rural flight, 170
rural school development, 76, 107,
　175–177, 183, 215*n*
Rural School Development Project
　(RSDP), 17, 18, 94, 100, 178
Russia, 7, 11–13, 26, 32, 34–38, 42–43,
　55–58, 63, 86–87, *91*, 93, 126,
　127, 128, 195, 208*n*. *See also*
　Soviet Union
Russian Federation, 73, 86, 162,
　163, 208*n*
Russian language, 70, 71, 159
Russian schools, 36

Sachs, Jeffrey, 86–87
Samoff, Joel, 99
Sattler, Elisabeth, 63
Save the Children UK, 82, 112, 158,
　161, 162, 175, 194, 209*n*, 214*n*
Schluss, Henning, 63
Schöne, Uta, 33, 49
School 2001, 18, 109, 110, 112, 118,
　147, 148, 162, 213*n*
School Act of 1963, 46, 48
Schriewer, Jürgen, 6–7, 37, 187
Scientific Committee, 40–41
scribes, schools for, 31–33
script reform, 42, 103
secondhand borrowing, 9–10
secular education, 25–27, 29, 31–37,
　40, 41, 206*n*
sedentarization, 13
Shagdar, S., 26, 33, 37, 42, 70
Shagdarsüren, L., 33, 46–47, 206*n*
shamanism, 17, 26–28
Sharkhüü, Ts., 31, 32
Shine Tol', 37
shadow economy, 92–93
Silarszky, P., 209*n*
Smith, J., 211*n*
Sneath, David, 13, 58, 64, 160, 170,
　171, 213*n*
social justice, 5, 6, 83
socio-logic, 7–8, 187
Soros Foundation. *See* Open Society
　Institute (Soros Foundation)
Soros Foundations Network, 82–83
South Africa, 124, 125, 133–135, 200
Soviet October Revolution of 1921,
　55–56

Soviet Union, 2, 7, 11, 26, 38–39, 41–45, 48–49, 60, 63, 67–69, 72–73, 78–79, 87, 89, 129, 156, 159, 188, 191, 195. *See also* Russia
Spreen, Carol Anne, 135
Stalin, Josef, 51, 59, 69
standardized tests/testing, 124, 132–133
Steiner-Khamsi, Gita, 17–18, 19, 24, 119, 144
Stolpe, Ines, 18
structural adjustment policies (SAPs), 85–86, 94, 102, 129, 134, 196–197
student-centered learning, 6, 14, 109–129, 133, 187, 196–197
Sum"yaasüren, Balchinbazar, 19, 118, 119

task monitors, 115–116
Teacher 2005, 18
Teacher as Parents study, 96, 98, 144
teacher salaries, 96–97, 138–141, 210n, 212n
teacher scorecards, 131, 133–134, 136, 139, 141, 143, 144, 198
teachers, discipline and punishment of, 137–141
teleology of transition, 63–64
Third International Mathematics and Science Study (TIMSS), 3–4, 124–125, 197, 211n, 215n
Third World countries, 6, 19, 60, 67, 71–72, 73, 78–78, 194, 196
Thomas, George, 125
Thomas, Helen T., 122
Tibet, 12, 25–29, 32, 34–35, 129, 188, 205n, 215n
Tilly, Charles, 23, 202
transition and transformation countries, 63–64
traveling reforms, 1, 8, 14, 129, 183, 193, 202
Truman, Harry, 215n
Tsagaan, R., 136
Tseveen, Jamsrany, 36, 37, 39, 40
tuition-based higher education, 94–99, 197, 212n
Tümendelger, Sengedorj, 19

Tümendemberel, Dagiisüren, 19
"two principles," 26–27, 32

Ulziisaikan, 208n
unemployment, 86, 87, 121, 125, 195, 211n, 214n
Ungern-Sternberg, baron, 38
uninterrupted education, 155–156
United Nations (UN), 35, 72, 74, 78–81, 189, 196
  Mongolia's entry to, 35, 79
universal access to education, 2, 16–17, 38–49, 88, 128, 172, 188
"urbanism," 13, 170
"urbanization," 13, 17, 153, 170, 174, 175
Urtnasan, N., 39
U.S.S.R. *See* Soviet Union

Verdery, Katherine, 15, 16, 65
Verheyen, G., 209n
Vietnam, 72, 207n
Vladimirtsov, B., 57
Voslensky, Michael S., 63
vouchers, 4, 18, 133, 135, 147–157, 159, 161–163, 182, 187, 196, 198, 199, 202
  use of the term in Mongolia, 148–149

Wallerstein, I., 24
Weidman, John, 209n
World Bank, 17, 74–78, 85–89, 93, 103, 144, 163, 180, 187, 189, 208n, 210n, 213n
world culture theory, 5–6, 126, 197–198, 202
world-systems, 6, 11, 23–24, 78, 79, 123–124, 128, 187, 188, 194–196
  modern world-system, 23–24
World Vision, 82, 158, 194
World War I, 23
World War II, 11, 35, 42, 67, 68, 127

Yellow Hats, 27, 205n
Yuan Dynasty, 25, 205n

zeitgeist, 25, 46

GPSR Compliance
The European Union's (EU) General Product Safety Regulation (GPSR) is a set of rules that requires consumer products to be safe and our obligations to ensure this.

If you have any concerns about our products, you can contact us on

ProductSafety@springernature.com

In case Publisher is established outside the EU, the EU authorized representative is:

Springer Nature Customer Service Center GmbH
Europaplatz 3
69115 Heidelberg, Germany

www.ingramcontent.com/pod-product-compliance
Lightning Source LLC
LaVergne TN
LVHW051915060526
838200LV00004B/161